Exterior Algebras

Series Editor
Nikolaos Limnios

Exterior Algebras

Elementary Tribute to Grassmann's Ideas

Vincent Pavan

iSTE PRESS

ELSEVIER

First published 2017 in Great Britain and the United States by ISTE Press Ltd and Elsevier Ltd

ISTE Press Ltd
27-37 St George's Road
London SW19 4EU
UK

www.iste.co.uk

Elsevier Ltd
The Boulevard, Langford Lane
Kidlington, Oxford, OX5 1GB
UK

www.elsevier.com

British Library Cataloguing-in-Publication Data
A CIP record for this book is available from the British Library
Library of Congress Cataloging in Publication Data
A catalog record for this book is available from the Library of Congress
ISBN 978-1-78548-237-3

Printed and bound in the UK and US

Contents

Preface

Hermann Grassmann. It is not that he has been unfairly criticized by his peers (as it was in the case of L. Boltzmann). Simply, even without exaggeration, they purely and solely ignored him. His ideas, dazzling compared to contemporary mathematical theories, are, however, the foundation of several modern algebraic constructions. Of course, it is an understatement to claim that Grassmann was too far ahead of his time. But how could it be possible to restrain oneself in front of such aberration that today, more than two hundred years after the birth of its author (1809), Grassmann's algebra is not included in basic education yet? And there is also this question, an answer to which is hardly conceivable: what is the reason for being forced to remain within the stifling space of vectorial formalism, which is inadequate for geometrical problems, whereas even the *multivector* formalism yields remarkably effective answers?

Without doubt, we might paraphrase the sociologist and leave it up to his remark, full of disarming cruelty: "*There is no intrinsic force in a true idea*". Although, in the most objective terms, Grassmann's algebra represents an ideal framework for mathematical discourse about numerous aspects, it will most likely remain unknown unless anybody is actually prompted to it. Incidentally, this issue clearly demystifies the creation of scientific knowledge, ideally glorified (by its own creators) as embodiment of triumphant reason, even though it so obviously obeys some logical discords, which, alas, do not always contribute to the useful spread of flourishing ideas.

There would be no end to it, if we tried to list all the domains of the utmost importance in which exterior algebra plays a crucial role: differential calculus, differential and projective geometry, electromagnetism, mechanics (classical, relativistic, quantum), etc. And yet, these are only the applications with the most evident dependence on exterior algebras. At a more basic level, it suffices to explain to anybody who seeks further arguments that Grassmann's algebra is simply the best possible framework for dealing with determinants. At the end of the day, what is most striking at a time when theoretical physics triumphs sensationally (how can we not be amazed by the discovery of gravitational waves), is the persistence of vector calculus tool in lieu of the multivector tool. Thus, bizarre objects such as the Levi-Civita symbol (of order 3, which some have been able to generalize for an order n) or, as we shall argue, the simple vector cross product are in truth no more than a rather peculiar and rather complex way of avoiding, without precisely visible reasons, the usage of Grassmann's formalism.

The worst of this systematically sustained and diffused ignorance is the fact that the construction of exterior algebra by means of solution of a universal problem is, today, a simple stylistic exercise that is entirely integrated within the algebraic discourse. While the basic concepts of linear algebra are routinely (and, doubtlessly, far too automatically) taught from the first steps in higher education, there is absolutely no difficulty, as soon as the vocabulary has been defined, in fully delving into the construction of tensor algebra, though it be skew-symmetric. Then, once the framework is set, structural questions are formulated *quite naturally* one after another, and the provided answers string together as simply as the questions that led to them.

Though, in order to keep this introduction completely honest, the reasons that inspired the writing should be addressed. A ridiculous story of a teacher, bewildered by his students' difficulties in understanding the formalism of Newtonian mechanics. It should be said that, sadly, we have reached a stage at which, in France, the succession of different education reforms (both in the secondary education and in the higher level education) has completely deprived the students of all serious means of doing even the slightest constructive science. Therefore, the teacher that cannot tackle directly three-dimensional mechanics anymore asks himself: how do we lay a formalism in two dimensions which could be conjugated in the same way in three dimensions (and, ultimately, in n dimensions)? However, in order for such an endeavor to be achieved, we should be able to rid ourselves of a cumbersome false friend: the vector cross product. A task which might seem impossible at first, as this algebraic operation structures the entire formalism of mechanics and electromagnetism. Yet, the vector cross product does have a generalization for any dimension: the Hodge conjugation of the exterior product. Then, the following conclusion is urged: the vector cross product can and should be replaced by the exterior product in every formalism of mechanics. Then, moments (of force or kinetic moments) should be considered as bivectors and not as vectors anymore. Then, it becomes obvious to move on to the construction of exterior algebra.

As P. Bourdieu used to say, "*science is done just as much with its education as against its education*". And it is precisely with this mindset that this book is written. For mechanics, but also definitely against it. This book is thus an algebra volume. But most of all, it is not a book of an algebraist. In particular, the writing will never try to avoid a detailed exposure, sometimes an extremely basic one, which professional algebraists will find very unnecessary. Instead, it is a guided tour into Grassmann's so well-posed world, with the ambition only to *shed light, as meticulously as possible, on the bases of exterior formalism*. This is a book which does not focus *a priori* on any specific use (although it is clearly impossible not to head towards some future uses) in order to leave open the field of possible uses. It is a genetically elementary book, which should be accessible to all those who have the need or the desire of becoming familiar with exterior algebra, be it aspiring physicists or computer specialists, mathematicians or students curious about exploring the dazzling world of algebraic constructions.

This work thus stems from a both modest and personal inspiration. Naturally, it borrows from many utterly scattered lectures, issued from radically different scientific environments, often unequal in quality and quantity, which can be found on the Internet. It would be impossible to list the whole set of sources, some of which consist of some few lines, based on which we have formulated the topics and have drawn the demonstrations illustrated in this document. Unfortunately, too few classical references on exterior algebras exist for us to reference. On condition of being able to separate the wheat from the chaff, which is never easy when dealing with information available on the Web, there is always something to retain from the documents found by means of search engines. Being a traditional edition, the only work that we consider as foundational is the one by L. Schwartz on *tensors* [SCH 81]. However, its presentation seems to be far too bourbakist to actually be able to spread outside of a sadly too limited audience. L. Schwartz himself has written in its autobiography (*A mathematician grappling with his century*) about tensors [SCH 97]: "*Some mathematicians are easily able to understand the tensor product of two vector spaces, but others have a lot of difficulty, and some quite reputable mathematicians have never succeeded in really using it*". We hope that this book will be more accessible and will help dissipate some misunderstandings.

Nevertheless, it would be illusory not to evoke the inevitable errors, imprecision, approximations, misprints or other clumsiness that necessarily have slipped into this book, escaping our vigilance in spite of all the precautions and feverish proofreading that we could achieve. We excuse ourselves in advance for any disagreement that this might cause, but all the same we remain confident regarding the pertinence of the work and the clarification that it will bring, whatever happens (as we hope), to exterior algebras. To all those who are about to follow the path of Grassmann's ideas, we bid you good reading.

Introduction

It is in 1844, at the age of 35, that Hermann Grassmann published his book "Die Lineale Ausdehnungslehre" [GRA 44][1]. Although it is well known today as a major work on algebra and geometry (an anthology of analyses and insightful comments on the relevance of Grassmann's ideas can be found in [SCH 96]), it seems that its contemporaries arrogantly ignored it, reproaching its author for employing a style way too incomprehensible to be studied. This was just the beginning of "*The Grassmann's tragedy*", as it was so pertinently called by J. Dieudonné [DIE 79]: "*In the whole gallery of prominent mathematicians who, since the time of the Greeks, have left their mark on science, Hermann Grassmann certainly stands out as the most exceptional in many respects, when compared with other mathematicians, his career is an uninterrupted succession of oddities: unusual were his studies; unusual his mathematical style; highly unusual in his own belated realization of his powers as a mathematician; unusual and unfortunate the total lack of understanding of his ideas, not only during his lifetime but long after his death; deplorable the neglect which compelled him to remain all his life professor in a high-school.*" And it was not only because Gibbs introduced his vector calculus, which became dominant in the formalism of physics, that Grassmann (and Clifford as well) sank into oblivion. If we believe Friedrich Engel [ENG 11] (as cited by Fearnaley-Sander [FEA 79]), Cauchy himself, who is known to have possessed a copy of Grassmann's work, exploited the ideas of "*Ausdehnungslehre*", without any allusion to their author, in a paper titled "*algebraic keys*" [CAU 82]. Even when the most famous mathematicians used Grassmann's ideas with a reference to him, they did so reluctantly. Thus, David Hestenes [PET 11], citing in his turn Engel [ENG 11] would write: "*Elie Cartan (1922) incorporated Grassmann's outer product into his calculus of differential forms. Though it put Grassmann's name into the mathematics mainstream, it so diluted his ideas that Engel called it "Cartanized Grassmann." In another irony,*

1 A French translation of this book exists, by D. Flament (and B. Bekemeier): *La Science de la grandeur extensive: la "lineale Ausdehnungslehre"*, Albert Blanchard Editeur, Paris, 1994.

Cartan (1968) also employed a matrix form of Clifford algebra in his "theory of spinors," but he failed to recognize its relation to Grassmann algebra and differential forms."

Although rewritten in 1862 in a form which Grassmann then considered to be more accessible [GRA 62], the ideas and the formalism originally contained in [GRA 44] found a widespread acceptance only in their original form, even if, as highlighted by D. Flament [FLA 05] at that point, in the 1870s, references to Grassmann on behalf of known mathematicians such as Klein, Hankel, or Lie, started emerging. According to Dieudonné [DIE 79], it is only because Cartan's formalism was better understood that Grassmann's ideas began to be reconsidered: "*only after 1930, when Elie Cartan's work started being understood, Grassmann's work regained its rightful central role in all the applications of algebra and multi-linear analysis*".

In this book, we introduce (at a level which we believe is accessible to most) the construction of exterior product and its basic usage, which has become so important and, actually, indispensable in mathematics and physics.

As D. Hestenes quite rightly said (see [SCH 96]): "*Still, in conception and applications, conventional renditions of his exterior algebra fall far short of Grassmann's original vision.*" This book, not more than others, is not an exception to this rule. It is simply a presentation, hopefully modern, of exterior products and its main usage within the formalism of linear and multi-linear algebra. In a certain way, it is a fulfillment (humble and rife with admiration for its inventor) of one of the wishes that its author expressed in the preface of his book: (quoted by [FEA 79]) "*I know that if I also fail to gather around me (as I have until now desired in vain) a circle of scholars, whom I could fructify with these ideas, and whom I could stimulate to develop and enrich them further, yet there will come a time when these ideas, perhaps in a new form, will arise anew and will enter a living communication with contemporary developments. For truth is eternal and divine.*" A living presentation therefore, employing everything that has been clarified by algebra over the years, with the purpose of sharing with the reader the fertile developments enabled by Grassmann's algebras. In order to guide the reader throughout this book, we briefly present here each chapter:

1) The first chapter contains some reminders of linear algebra. These are, essentially, some fundamental notions of infinite dimensional vector space, as those taught during the first years of high school. This chapter can be easily omitted by the most seasoned readers, who will not find anything new in it. Far from being a complete, autonomous course on vector spaces (in their most abstract aspects), this chapter provides just the right amount of foundation for algebraic structures that will make possible the construction of exterior afterwards. Algebraic bases are recalled, as well as some notions of linear application and of duality, equivalence relations and quotient spaces, permutations, multi-linearity and skew symmetry.

2) The second chapter *constructs* the exterior algebras by solving a universal problem. At first, it focuses on exterior algebras of degree 2, and then it explains how the considered construction can be generalized for algebras of degree p. It should be noted that this construction is achieved in a direct way, without first employing the more generic construction of tensors (although such an approach is highlighted at the end of the chapter). In the second section, uniqueness (up to isomorphism) of exterior algebras of degree p is shown. The chapter ends with some remarks and some vocabulary. The advantage of this chapter is that it avoids an axiomatic presentation of exterior algebras.

3) The third chapter analyzes in detail the construction of exterior product between elements of exterior algebras. It shows the basic technical properties of this operation (skew symmetry, distributivity, etc.) and makes it possible to define Grassmann's algebra, which is introduced as a graded algebra. Next, the exterior product between linear forms is constructed "by hand" and we employ this tool to demonstrate the most essential property of the exterior product between vectors: its nullity allows us to characterize its dependence.

4) The fourth chapter focuses on establishing the bases of exterior algebras. This construction is done when the original vector space has an algebraic basis which is indexed by a totally ordered set. In this situation, the case of finite dimension is detailed. Afterwards, we consider the characterization of dual spaces of exterior algebras. In particular, in the finite dimension case, the characteristic construction of skew-symmetric multi-linear forms is obtained thanks to exterior products of the forms, which are obtained in their turn as elements of the dual space of the exterior algebras.

5) The fifth chapter is dedicated to the topic of determinants. The latter are introduced naturally as components of p-vectors decomposed over bases of exterior products. By using this definition, classical properties of determinants are re-demonstrated and the usual methods of determinants calculation are reviewed. Thereof stems an observation that the exterior algebra approach for determinants is both effective and elegant, and it provides, for example, a rather ingenious way (although, actually, utterly methodical in its structure) of calculating "by hand" determinants of order 4 without any effort. It is also shown that exterior formalism, thanks to its nature intimately related to determinant calculus, enables the analysis of linear systems in a formal and effective way: in fact, Cramer formulas are easily re-demonstrated, while discovering a favorable framework for the enunciation of Rouché-Fontené theorem.

6) The sixth chapter is independent, in principle, from the exterior subject. It draws attention to the notion of pseudo-scalar product by recalling the analogies and differences in comparison to the (Euclidean) scalar product. These considerations focus on the definition of orthogonality, the characterization of nondegenerate subspaces, as well as on the construction of (pseudo-) orthonormal bases. The chapter ends with the classical Sylvester's law of inertia.

7) The seventh chapter swiftly introduces the pseudo-scalar product from a vector space to its exterior algebras. Thus, it becomes possible to construct pseudo-orthonormal bases of exterior algebras from pseudo-orthonormal bases of the vector space and to provide a characterization of non-degeneration. In the Euclidean case, the norm of the exterior product allows us to calculate the volume of a parallelotope or that of a simplex constructed upon the vectors in question. Thanks to an exterior calculation, an alternative to the calculation of Riemann measure of differential varieties.

8) The eighth chapter tackles the subject of divisibility and decomposability of elements in exterior algebras. As a preliminary phase to the study of these topics, the fundamental notion of contraction product is introduced as a dual of the exterior product. Thus, a fundamental theorem yields the characterization of the divisibility by anisotropic vectors. The last result of the chapter makes it possible to show a systematic result of decomposition in algebras of order $n - 1$.

9) The ninth chapter defines and studies the Hodge conjugation. At first, it is defined by means of contraction product. Next, the theorem about the divisibility by vectors from the previous chapter is expanded to include k-blades. Then, the relation between the Hodge conjugation and the vector division problem is shown. Consequently, we explain how, by using the Hodge conjugation, the notion of vector cross product can be generalized for any dimension. In the second section of the chapter, we define the regressive product as the conjugate of the exterior product, short of a conjugation. We deduce some properties of this operation, which is a useful tool in writing and calculations.

10) The tenth chapter raises the important question of endomorphisms in exterior algebra. We illustrate how it is possible to construct endomorphisms of exterior algebras starting from an endomorphism of a vector space. The notion of invariance of an endomorphism family is introduced, which helps address the topic of endomorphism invariants and their fundamental properties in a simple way. Afterwards, we quickly approach the notion of conjugated endomorphism. This notion allows us to tackle the subject of decomposability of endomorphisms of exterior algebras, which can finally be related to Laplace's inversion formula of isomorphisms.

11) The eleventh and final chapter focuses on a particular study of the exterior algebra of order 2. A fundamental result shows that this algebra is isomorphic with regard to skew-symmetric linear operators on the vector space. By studying the results of decomposability of 2-exterior algebra, the results of (pseudo-) orthogonal decomposition over (pseudo-) skew-symmetric operators are shown. Incidentally, the classical result of Cartan's lemma is demonstrated.

1

Reminders on Linear Algebra

In this section, we recall only the essential definitions and results of linear algebra which are useful for the rest of the book. We shall usually choose to consider the case of infinite dimension, which will be indispensable from the perspective of construction of exterior algebras in the next chapter. Most demonstrations are omitted to the extent that they are either quite simple or inappropriate for a chapter without the aim of repeating a detailed course on algebra.

1.1. Linear spaces

1.1.1. *Definition*

In all that follows, we shall limit ourselves to real linear spaces, that is, to set $\boldsymbol{E}$ equipped with a commutative group law + and an external product law (scalar multiplication) $\mathbb{R} \times \boldsymbol{E} \mapsto \boldsymbol{E}$, in order that the following axioms are verified:

1) Distributivity of scalar multiplication with respect to vector addition: $\lambda(\mathbf{x}+\mathbf{y}) = \lambda\mathbf{x} + \lambda\mathbf{y}$.

2) Distributivity of scalar multiplication with respect to addition over the reals: $(\alpha+\beta)\mathbf{x} = \alpha\mathbf{x} + \beta\mathbf{x}$.

3) Compatibility of scalar multiplication with multiplication over the reals: $(\alpha\beta)\mathbf{x} = \alpha(\beta\mathbf{x})$.

4) Identity element of scalar multiplication: $1\mathbf{x} = \mathbf{x}$.

Although we do not seek to repeat a course on linear spaces, in a first phase we recall some definitions and some properties of infinite dimensional spaces, which are often less known than their counterparts with finite dimension, since the construction of exterior algebras initially considers such spaces. A linear subspace of $\boldsymbol{E}$ is any subset $\boldsymbol{F} \subset \boldsymbol{E}$ such that:

1) It is non-empty: $\boldsymbol{F} \neq \varnothing$.

2) It is stable with respect to addition: $\mathbf{x}, \mathbf{y} \in \boldsymbol{F} \Rightarrow \mathbf{x} + \mathbf{y} \in \boldsymbol{F}$.

3) It is stable with respect to scalar multiplication: $\lambda \in \mathbb{R}, \mathbf{x} \in \boldsymbol{F} \Rightarrow \lambda\mathbf{x} \in \boldsymbol{F}$.

Finally, for any non-empty subset $F \subset \boldsymbol{E}$, there exists a unique smallest linear subspace of $\boldsymbol{E}$ (in the sense of inclusion) that contains F. This subspace is called the subspace generated by F and it is denoted by $\langle F \rangle$.

1.1.2. *Algebraic basis of linear spaces*

DEFINITION 1.1.– Let $\boldsymbol{E}$ be a linear space.

1) A set $\mathbf{e}_i, i \in I$ is called a generating set of $\boldsymbol{E}$ if for every $\mathbf{x}$ of $\boldsymbol{E}$ there exists a finite subset $J(\mathbf{x}) \subset I$ of elements of I, as well as a set of reals $x_j, j \in J(\mathbf{x})$, such that it can be written as:

$$\mathbf{x} = \sum_{j \in J(\mathbf{x})} x_j \mathbf{e}_j. \quad [1.1]$$

2) A set $\mathbf{e}_i, i \in I$ is said to be a linear independent family in $\boldsymbol{E}$ if for every finite subset $J \subset I$ of elements of I and for any set $\lambda_j, j \in J$ the following implication ensues:

$$\sum_{j \in J} \lambda_j \mathbf{e}_j = \mathbf{0} \Rightarrow \forall j \in J,\ \lambda_j = 0. \quad [1.2]$$

3) A set which is a linear independent family and a generating set of $\boldsymbol{E}$ is called an algebraic basis of $\boldsymbol{E}$.

4) A linear space which admits a basis with a finite cardinality is finite-dimensional. In such case, all bases of the space have the same cardinality, which is called dimension of the linear space.

The definition of algebraic bases is important when infinite dimensional linear spaces are considered. Attention should be paid in order not to mistake the algebraic bases with Hilbert bases: both are indeed bases of a linear space with a non-finite dimension; however, they do not have much in common. In particular, Hilbert bases suppose a countable cardinality. A fundamental theorem of linear algebra assures that every linear space admits a basis, i.e. the incomplete basis theorem.

THEOREM 1.1.– Let $\boldsymbol{E}$ be a linear space, then if $\mathcal{L}$ is a linearly independent family of $\boldsymbol{E}$ and $\mathcal{G}$ is a generating set of $\boldsymbol{E}$ such that $\mathcal{L} \subset \mathcal{G}$, there exists a basis $\mathcal{B}$ of $\boldsymbol{E}$ such that $\mathcal{L} \subset \mathcal{B} \subset \mathcal{G}$.

PROOF.– We admit the proof of this theorem which essentially is based on the application of Zorn lemma.

There is an immediate corollary to this result.

COROLLARY 1.1.– The following propositions are true:

1) From any generating set of $\boldsymbol{E}$ it is possible to extract an algebraic basis;

2) Every linearly independent family of $\boldsymbol{E}$ can be completed to obtain an algebraic basis of $\boldsymbol{E}$.

1.2. Linear applications

1.2.1. *Definition*

We recall that a linear map from $\boldsymbol{E} \mapsto \boldsymbol{F}$, where $\boldsymbol{E}, \boldsymbol{F}$ are two linear spaces, is a morphism which preserves the structure of the linear space: the image of a sum is the sum of the images, the image of a scalar multiplication is the product between the scalar and the image. It is possible to resume this with the following definition.

DEFINITION 1.2.– Let $\boldsymbol{E},\boldsymbol{F}$ be two linear spaces. An application $L : \boldsymbol{E} \mapsto \boldsymbol{F}$ is linear when:

$$\forall \mathbf{x}, \mathbf{y} \in \boldsymbol{E},\ \forall \lambda \in \mathbb{R},\ L(\mathbf{x} + \lambda \mathbf{y}) = L(\mathbf{x}) + \lambda L(\mathbf{y}). \quad [1.3]$$

The set of linear maps from $\boldsymbol{E}$ to $\boldsymbol{F}$ also has the structure of a linear space.

PROPOSITION 1.1.– Let $\boldsymbol{E}, \boldsymbol{F}$ be two linear spaces. We denote by $\mathcal{L}(\boldsymbol{E}, \boldsymbol{F})$ the set of linear maps defined over $\boldsymbol{E}$ with values in $\boldsymbol{F}$. It is a linear space for the pointwise addition and scalar multiplication. When $\boldsymbol{F} = \mathbb{R}$, we talk about a space of linear forms over $\boldsymbol{E}$, or in other terms of a (algebraic) dual space of $\boldsymbol{E}$, and we denote it by $\mathcal{L}(\boldsymbol{E}, \mathbb{R}) = \boldsymbol{E}^*$.

PROOF.– It is left to the reader to verify this immediate demonstration (regarding the linear space structure of $\mathcal{L}(\boldsymbol{E}, \boldsymbol{F})$).

REMARK 1.1.– Let us note that the algebraic dual is different from "another" dual, the topological dual space, in the sense that there is no topology in the considered linear spaces ***E***, ***F***. ■

1.2.2. *Injections - surjections - identification*

PROPOSITION 1.2.– A linear map $L : \boldsymbol{E} \mapsto \boldsymbol{F}$ is surjective if and only if the image of an algebraic basis of ***E*** is a generating set of ***F***.

PROOF.– Let $\mathbf{y} \in \boldsymbol{F}$. By surjectivity, there exists $\mathbf{x} \in \boldsymbol{E}$ such that $\mathbf{y} = L(\mathbf{x})$. Let then $\mathbf{e}_i, i \in I$ be a basis of ***E***. It is known that there exists a finite $J(\mathbf{x}) \subset I$ such that $\mathbf{x} = \sum_{j \in J(\mathbf{x})} x_j \mathbf{e}_j$. Thus, we conclude that:

$$\mathbf{y} = L(\mathbf{x}) = L\left(\sum_{j \in J(x)} x_j \mathbf{e}_j\right) = \sum_{j \in J(\mathbf{x})} x_j L(\mathbf{e}_j), \qquad [1.4]$$

which actually shows that the family $L(\mathbf{e}_i), i \in I$ is a generating set of ***F***. *Inversely*, let us suppose that the image of every basis is a generating set. Let $\mathbf{y} \in \boldsymbol{F}$. Since $L(\mathbf{e}_i), i \in J$ is a generating set, there exists $J(\mathbf{y})$, a finite subset of I, as well as a family $y_j, j \in J(\mathbf{y})$ such that:

$$\mathbf{y} = \sum_{j \in J(\mathbf{y})} y_j L(\mathbf{e}_j) = L\left(\sum_{j \in J(\mathbf{y})} y_j \mathbf{e}_j\right). \qquad [1.5]$$

Then, the element $\mathbf{x} = \sum_{j \in J(\mathbf{y})} y_j \mathbf{e}_j$ satisfies $L(\mathbf{x}) = \mathbf{y}$, which proves the surjectivity of the application L

Let us continue with the related proposition regarding the characterization of injective applications:

PROPOSITION 1.3.– A linear map $L : \boldsymbol{E} \mapsto \boldsymbol{F}$ is injective if and only if the image of an algebraic basis of ***E*** is a linearly independent set of ***F***.

PROOF.– Suppose that L is injective and consider a zero combination (with finite support) of the family $L(\mathbf{e}_i), i \in I$. We have:

$$\sum_{j \in J} \lambda_j L(\mathbf{e}_j) = 0 \Leftrightarrow L\left(\sum_{j \in J} \lambda_j \mathbf{e}_j\right) = 0 \Leftrightarrow \sum_{j \in J} \lambda_j \mathbf{e}_j = 0, \qquad [1.6]$$

the latter equivalence being obtained from the fact that L is injective. Now, the last equality to 0 can happen only if $\forall j \in J, \lambda_j = 0$, since the family of $\mathbf{e}_i, i \in I$ is an algebraic basis. Therefore, the family $L(\mathbf{e}_i), i \in I$ is indeed a linearly independent set. *Inversely*, suppose that the image of every algebraic basis is a linearly independent set. Let us solve $L(\mathbf{x}) = 0$. It is always possible to write $\mathbf{x}$ as a linear combination (with finite support) of $\mathbf{e}_i, i \in I$. We have then:

$$L\left(\sum_{j \in J} x_j \mathbf{e}_j\right) = 0 \Leftrightarrow \sum_{j \in J} x_j L(\mathbf{e}_j) = 0. \qquad [1.7]$$

Now, as the family $L(\mathbf{e}_i)$, $i \in I$ is a linearly independent set, this means that every x_i is null and therefore that $\mathbf{x} = 0$. Hence, the injectivity of L.

Finally, a uniqueness theorem (and thus a theorem of identification) of a linear map:

PROPOSITION 1.4.– Let $\boldsymbol{E}$ and $\boldsymbol{F}$ be two linear spaces. Let $\mathbf{e}_i, i \in I$ be an algebraic basis of $\boldsymbol{E}$ and $\mathbf{a}_i, i \in I$ a family of $\boldsymbol{F}$. Then, there exists a unique linear map $L : \boldsymbol{E} \mapsto \boldsymbol{F}$ such that:

$$\forall i \in I, \;\; L(\mathbf{e}_i) = \mathbf{a}_i. \qquad [1.8]$$

PROOF.– Regarding existence, it is sufficient to consider the application L defined by:

$$\mathbf{x} = \sum_{j \in J(x)} x_j \mathbf{e}_j \Rightarrow L(\mathbf{x}) := \sum_{j \in J(\mathbf{x})} x_j \mathbf{a}_j. \qquad [1.9]$$

Regarding uniqueness, it is sufficient to see that:

$$\begin{aligned} &\forall i \in I, \;\; L(\mathbf{e}_i) = L'(\mathbf{e}_i) = \mathbf{a}_i \\ &\qquad \Rightarrow \forall \mathbf{x} = \sum_{j \in J(x)} x_j \mathbf{e}_j, \; L(\mathbf{x}) = L'(\mathbf{x}) = \sum_{j \in J(x)} x_j \mathbf{a}_j. \end{aligned} \qquad [1.10]$$

1.2.3. *Algebraic duality*

We have defined the (algebraic) duality as a set of linear forms over $\boldsymbol{E}$. We shall now see how this space can be related to the linear space $\boldsymbol{E}$ itself.

PROPOSITION 1.5.– Let $\boldsymbol{E}$ be a linear space. Then, there exists an injection in its dual $\boldsymbol{E}^*$: in other words, there exists an injective linear map $\mathcal{L} : \boldsymbol{E} \mapsto \boldsymbol{E}^*$ which can be constructed in the following way: we associate with any basis $\mathbf{e}_i, i \in I$ the dual family $\phi_j^*, j \in I$ such that:

$$\forall i, j \in I, \;\; \phi_j^* \left(\mathbf{e}_i\right) = \delta\left(i, j\right), \qquad [1.11]$$

where $\delta\left(i,j\right)$ is the Kronecker symbol over I, i.e. $\delta\left(i,j\right)$ is null except for $i = j$, where it equals 1. Then, the family $\phi_i^*, i \in I$ is a linearly independent set of $\boldsymbol{E}^*$. Furthermore, when I has a finite cardinality (in other words, when $\boldsymbol{E}$ is finite-dimensional), the family $\phi_i^*, i \in I$ is also a generating set.

PROOF.– It is clear that, given a basis of $\boldsymbol{E}$, the construction of the family ϕ_j^*, $j \in I$ by means of formula [1.11] is possible in one unique way: this is an application of proposition 1.4. This family is obviously a linearly independent set, since if we suppose that for a finite subset $J \subset I$ we have:

$$\sum_{j \in J} \alpha_j \phi_j^* \left(\cdot\right) = 0\left(\cdot\right), \qquad [1.12]$$

then this means that for all $\mathbf{e}_i, i \in J$, we have:

$$\sum_{j \in J} \alpha_j \phi_j^* \left(\mathbf{e}_i\right) = 0 \Leftrightarrow \sum_{j \in J} \alpha_j \delta\left(i,j\right) = 0 \Leftrightarrow \alpha_i = 0. \qquad [1.13]$$

Therefore, the family $\phi_j^*, j \in I$ is indeed a linearly independent set, which shows the injectivity of the application $\mathbf{e}_i \mapsto \phi_i^*$ (see proposition 1.3). Suppose now that I has a finite cardinality. Then, we can set $I = [1, n]$. Then, the family $\phi_i^*, i \in [1, n]$ is also a generating set. In fact, let us consider a linear form ϕ^* over $\boldsymbol{E}$. For the basis $\mathbf{e}_i, i \in [1, n]$, we naturally set that:

$$\phi^* \left(\mathbf{e}_i\right) = \eta_i \textit{ and thus } \mathbf{x} = \sum_{j=1}^{j=n} x_j \mathbf{e}_j$$

$$\Rightarrow \phi^* \left(\mathbf{x}\right) = \sum_{j=1}^{j=n} x_j \phi^* \left(\mathbf{e}_j\right) = \sum_{j=1}^{j=n} x_j \eta_j = \sum_{j=1}^{j=n} x_j \sum_{i=1}^{i=n} \eta_i \phi_i^* \left(\mathbf{e}_j\right). \qquad [1.14]$$

By employing the linearity of applications ϕ_i^*, we can write:

$$\forall \mathbf{x} \in \boldsymbol{E}, \;\; \phi^* \left(\mathbf{x}\right) = \sum_{i=1}^{i=n} \eta_i \phi_i^* \left(\sum_{j=1}^{j=n} x_j \mathbf{e}_j\right) = \sum_{i=1}^{i=n} \eta_i \phi_i^* \left(\mathbf{x}\right), \qquad [1.15]$$

which is sufficient to show the generating nature of the family $\phi_i^*, i \in [1, n]$. Therefore, in the case of the finite dimension, we possess a method that allows us to transform a basis of $\boldsymbol{E}$ into a basis of $\boldsymbol{E}^*$, thus showing the isomorphism between the two spaces.

REMARK 1.2.– Some remarks regarding this proposition:

1) In the case of an infinite dimensional linear space, the space $\boldsymbol{E}$ is *never* isomorphic to its dual $\boldsymbol{E}^*$: this is a consequence of the Erdös-Kaplansky theorem which can identify the dual of $\boldsymbol{E}$. We have indeed an injection of $\boldsymbol{E}$ into $\boldsymbol{E}^*$ (according to construction 1.11), but we have no surjection.

2) The case of Hilbert or Hermitian spaces is a separate one, in the sense that the notion of associated basis is different from the definition of algebraic bases. ■

1.2.4. *Space of functions zero everywhere except in a finite number of points*

An important example of linear space and algebraic basis, which will be useful in what follows, is the following:

PROPOSITION 1.6.– Let E be any set. Then, the set $\mathcal{F}_0(E \mapsto \mathbb{R})$ of applications from E to $\mathbb{R}$ *that are zero everywhere except in a finite number of points in* E is a linear space for the pointwise addition and scalar multiplication. For $x \in E$, we consider the application $\delta_x(y)$ which equals 1 if $y = x$ and 0 otherwise. Then:

$$\textit{the family } \delta_x(\cdot), \ x \in E \ \textit{ is an algebraic basis of } \mathcal{F}_0(E \mapsto \mathbb{R}). \qquad [1.16]$$

PROOF.– First, it is clear that each application $\delta_x(\cdot)$ is actually zero everywhere except in a (finite number of) point and thus it is an element of $\mathcal{F}_0(E \mapsto \mathbb{R})$. We show at first that the considered family is a linearly independent set. Let $J \subset E$ be a finite non-empty subset of E. We denote by $x_1, \cdots, x_n$ the elements (distinct two by two) of J. Suppose that we have:

$$\sum_{i=1}^{i=n} \lambda_i \delta_{x_i}(\cdot) = 0 \Leftrightarrow \forall y \in E, \ \sum_{i=1}^{i=n} \lambda_i \delta_{x_i}(y) = 0. \qquad [1.17]$$

This means, by substituting y with each x_j:

$$\forall j \in [1, n], \ \sum_{i=1}^{i=n} \lambda_i \delta_{x_i}(x_j) = 0$$

$$\Leftrightarrow \forall j \in [1, n], \ \sum_{i=1}^{i=n} \lambda_i \delta(i, j) = 0 \Leftrightarrow \forall j \in [1, n], \ \lambda_j = 0, \qquad [1.18]$$

where $\delta(i,j)$ is the Kronecker symbol: it equals 1 if $i = j$ and 0 otherwise. Let us show that the family is a generating set. Let $f \in \mathcal{F}_0(E \mapsto \mathbb{R})$. We denote by $x_1, \cdots, x_n$ the set of points of E where f is non-zero. It follows immediately that:

$$\forall y \in E,\ f(y) = \sum_{i=1}^{i=n} f(x_i)\,\delta_{x_i}(y), \qquad [1.19]$$

which ends the demonstration.

1.3. Partition, quotient linear space

1.3.1. *Partition and equivalence relation*

The partition of a set makes it possible to sort all its elements into a list of subsets, so that each element belongs uniquely to a subset:

DEFINITION 1.3.– Let E be a set and $\mathcal{P}(E)$ be the set of subsets of E. We call partition of E a family $\theta_i, i \in I$ of subsets of E such that the three following properties hold:

$$\begin{array}{l} 1)\ \forall i \in I,\ \theta_i \neq \varnothing. \\ 2)\ \bigcup_{i\in I} \theta_i = E. \\ 3)\ \forall i,j \in I,\ i \neq j \Rightarrow \theta_i \cap \theta_j = \varnothing. \end{array} \qquad [1.20]$$

The existence of a partition helps consider the relation between two elements of a set in a "natural" way: two elements will be said to be in relation, more precisely "equivalent" if they belong to the same subset of the partition. This enables us to define a graph over E, i.e. a subset of $E \times E$, in the following way:

DEFINITION 1.4.– Let $\theta_i, i \in I$ be a partition of E. We call graph G associated with the partition of the subset of $E \times E$ defined by:

$$(x,y) \in G \Leftrightarrow \exists i \in I,\ [x \in \theta_i \ \textit{et}\ y \in \theta_i]. \qquad [1.21]$$

A list of properties stems from this definition:

PROPOSITION 1.7.– Graph G associated with a partition has the following properties:

1) For any x in E, we have $(x, x) \in G$;

2) For any x and y in E, we have $(x, y) \in G \Rightarrow (y, x) \in G$;

3) For any x,y and z in E, we have: $(x, y) \in G$ and $(y, z) \in G$ entails $(x, z) \in G$.

Before approaching the demonstration of this proposition, let us give some vocabulary:

1) The first property is called reflexivity of the graph.

2) The second property is called symmetry of the graph.

3) The third property is called transitivity of the graph.

PROOF.– Let us demonstrate these properties:

1) Let $x \in E$, then since $\bigcup_{i \in I} \theta_i = E$, it means that by definition there exists an element $i \in I$ such that $x \in \theta_i$. We can then write clearly that the proposition $\exists i \in I,\ [x \in \theta_i \textit{ and } x \in \theta_i]$ is true. Therefore, we have indeed $(x, x) \in G$.

2) Suppose that $(x, y) \in G$ which means that $\exists i \in I,\ [x \in \theta_i \textit{ and } y \in \theta_i]$. Of course, this implies that by "symmetry" of the symbol "and" that we have $\exists i \in I,\ [y \in \theta_i \textit{ and } x \in \theta_i]$, and thus we have indeed $(y, x) \in G$.

3) Finally, suppose that $\exists i \in I,\ [y \in \theta_i \textit{ and } x \in \theta_i]$ and $\exists j \in I,\ [y \in \theta_j \textit{ and } z \in \theta_j]$. It can then be deduced that the element y is both in θ_i and in θ_j. This implies that $\theta_i \bigcap \theta_j \neq \varnothing$ and thus necessarily that $i = j$. Therefore, we have the existence of an i such that x, y, z are all in θ_i and thus x, z are in the same θ_i, which implies that (x, z) is in G.

We have just seen that the graph associated with a partition satisfies the three properties above. We shall now demonstrate the inverse:

PROPOSITION 1.8.– Let G be a graph of $E \times E$ that satisfies the three following properties:

1) Reflexivity: $\forall x \in E,\ (x, x) \in G$;

2) Symmetry: $\forall x, y \in E,\ (x, y) \in G \Rightarrow (y, x) \in G$;

3) Transitivity: $\forall x, y, z \in E,\ (x, y) \in G \textit{ et } (y, z) \in G \Rightarrow (x, z) \in G$.

Then, we can construct a partition $\theta_i, i \in I$ over E so that G is exactly the graph associated with this partition.

Before continuing with the actual demonstration of this result, we shall introduce some vocabulary:

DEFINITION 1.5.– Let E be a set and G a graph over E, i.e. a subset of $E \times E$. For any element x in E, we call class of x, denoted by $cl\left(x\right)$ the set of all the elements y in E such that (x, y) is an element of the graph:

$$\forall x \in E, \ \ cl\left(x\right) = \{y \in E, \ (x,y) \in G\}. \qquad [1.22]$$

We denote by $\mathcal{C}$ the set of all the classes:

$$\mathcal{C} = \{cl\left(x\right), \ x \in E\} \Leftrightarrow [\mathfrak{c} \in \mathcal{C} \Leftrightarrow \exists z \in E, \ \ \mathfrak{c} = cl\left(z\right)]. \qquad [1.23]$$

Now, let us continue with the proof of the proposition: in fact, we shall show that the set of classes forms a partition of E.

PROOF.– Let us show in sequence the three properties which allow us to define the partitions:

1) First, it is clear, by definition, that each element of $\mathcal{C}$ represents a non-empty subspace of E. In fact, if $\mathfrak{c} \in \mathcal{C}$, then there exists z such that $\mathfrak{c} = \{x, (z, x) \in G\}$. Now, since the graph is reflexive, we know that $\forall z \in E, (z, z) \in G$, and thus we have indeed $z \in \mathfrak{c}$ which therefore is non-empty.

2) Then, we can see that $\mathfrak{c} \neq \mathfrak{d} \Leftrightarrow \mathfrak{c} \bigcap \mathfrak{d} = \varnothing$. In order to do this, we shall rather show that we have $\mathfrak{c} = \mathfrak{d} \Leftrightarrow \mathfrak{c} \bigcap \mathfrak{d} \neq \varnothing$ which is equivalent. Now, it is indeed true that $\mathfrak{c} = \mathfrak{d}$ if and only if $\exists\, x \in \mathfrak{c}, \ \exists y \in \mathfrak{d}$ such that $(x, y) \in G$. Suppose, in fact, that $x \in \mathfrak{c} \bigcap \mathfrak{d}$. Then, by transitivity and by symmetry of the graph, we deduce that any y in $\mathfrak{c}$ is also in $\mathfrak{d}$ and vice versa, and thus that $\mathfrak{c} = \mathfrak{d}$. *Inversely*, if $\mathfrak{c} = \mathfrak{d}$, as $\mathfrak{c}$ and $\mathfrak{d}$ are non-empty, it is clear that there exists x in the intersection.

3) Finally, since we have:

$$\forall x \in E, \ (x, x) \in G \Rightarrow \forall x \in E, \ x \in cl\left(x\right) \Rightarrow$$
$$E \subset \bigcup_{x \in E} cl\left(x\right) \Rightarrow E = \bigcup_{x \in E} cl\left(x\right), \qquad [1.24]$$

since the inclusion $\bigcup_{x \in E} cl\left(x\right) \subset E$ is always guaranteed.

We deduce that the family $\mathfrak{c}, \mathfrak{c} \in \mathcal{C}$ is indeed a partition of the set E. By its construction, such partition is indeed the one which is associated with the graph G.

Therefore, we see that there is an exact correspondence between the partitions and the equivalence relations. In fact, the equivalence relation is a very powerful tool for constructing partitions. Thus, we most often use the equivalence of points encountered while identifying the equivalence relations for partitions construction.

1.3.2. *Quotient linear space*

DEFINITION 1.6.– Let E be a set and G a graph over E.

1) We define over $E \times E$ the symbol of relation R according to the following formula:

$$\forall x \in E,\ \forall y \in E,\ x\mathrm{R}y \Leftrightarrow (x, y) \in G. \qquad [1.25]$$

2) A relation over E is called an equivalence relation when the graph G satisfies the properties of proposition 1.7.

3) The classes associated with an equivalence relation are called equivalence classes.

4) The set of all equivalence classes associated with an equivalence relation is called the quotient set. It is denoted by E/R.

5) The application $x \in E \mapsto cl(x) \in E/\mathrm{R}$ which associates with an element its equivalence class is called the canonical surjection.

EXAMPLE 1.1.– Let $\boldsymbol{F}$ be a linear subspace of a linear space $\boldsymbol{E}$. We can always write construct a relation R in $\boldsymbol{E}$ according to:

$$\forall \mathbf{x}, \mathbf{y} \in \boldsymbol{E},\ \mathbf{x}\mathrm{R}\mathbf{y} \Leftrightarrow \mathbf{x} - \mathbf{y} \in \boldsymbol{F}. \qquad [1.26]$$

It is clear that this relation is an equivalence relation, because the linear subspaces:

1) contain the null vector (which implies reflexivity);

2) are stable with respect to addition (which implies transitivity);

3) are stable with respect to scalar multiplication (and if the scalar equals -1, it entails symmetry).

We can then define the quotient set $\boldsymbol{E}/\mathrm{R}$ which is denoted by $\boldsymbol{E}/\boldsymbol{F}$. We say that we quotient $\boldsymbol{E}$ by $\boldsymbol{F}$. ■

Let us introduce the following definition:

DEFINITION 1.7.– We say that an equivalence relation $\mathtt{R}$ defined over a linear space $\boldsymbol{E}$ is compatible with the linear space operations when the two following properties are satisfied:

1) $\forall \mathbf{y}, \mathbf{x} \in \boldsymbol{E}, \forall \lambda \in \mathbb{R}, \mathbf{x}\mathtt{R}\mathbf{y} \Rightarrow (\lambda\mathbf{x})\,\mathtt{R}\,(\lambda\mathbf{y})$.

2) $\forall \mathbf{x}, \mathbf{y}\mathbf{x}', \mathbf{y}' \in \boldsymbol{E}$, $\mathbf{x}\mathtt{R}\mathbf{y}$ *and* $\mathbf{x}'\mathtt{R}\mathbf{y}' \Rightarrow (\mathbf{x}+\mathbf{x}')\,\mathtt{R}\,(\mathbf{y}+\mathbf{y}')$.

EXAMPLE 1.2.– The quotient operation of linear subspaces is always compatible with the linear space operations. In fact:

1) Suppose that $\mathbf{x}\mathtt{R}\mathbf{y}$ and $\lambda \in \mathbb{R}$, then we have $(\mathbf{x}-\mathbf{y}) \in \boldsymbol{F}$. Since $\boldsymbol{F}$ is a linear subspace, then $\lambda(\mathbf{x}-\mathbf{y}) = \lambda\mathbf{x} - \lambda\mathbf{y}$ is still in $\boldsymbol{F}$ and thus $\lambda\mathbf{x}\mathtt{R}\lambda\mathbf{y}$;

2) Suppose now that $\mathbf{x}\mathtt{R}\mathbf{y}$ and $\mathbf{x}'\mathtt{R}\mathbf{y}'$. Then, we know that $(\mathbf{x}-\mathbf{y}) \in \boldsymbol{F}$ and $(\mathbf{x}'-\mathbf{y}') \in \boldsymbol{F}$. According to the stability of the sum over linear subspaces, we have $(\mathbf{x}-\mathbf{y}) + (\mathbf{x}'-\mathbf{y}') \in \boldsymbol{F}$, i.e. precisely $(\mathbf{x}+\mathbf{x}') - (\mathbf{y}+\mathbf{y}') \in \boldsymbol{F}$, which means exactly $(\mathbf{x}+\mathbf{x}')\,\mathtt{R}\,(\mathbf{y}+\mathbf{y}')$. ■

The equivalence relations compatible with linear space operations thus make it possible to construct new linear spaces, as is shown by the following theorem:

THEOREM 1.2.– Let $\boldsymbol{E}$ be a linear space and $\boldsymbol{F}$ a linear subspace of $\boldsymbol{E}$. We denote by $\boldsymbol{E}/\boldsymbol{F}$ the quotient space of $\boldsymbol{E}$ by $\boldsymbol{F}$ and by cl the canonical surjection of $\boldsymbol{E}$ in $\boldsymbol{E}/\boldsymbol{F}$. Then, we can equip the quotient space with a linear space structure by setting:

$$cl(\mathbf{x}) + cl(\mathbf{y}) := cl(\mathbf{x}+\mathbf{y}), \quad \lambda cl(\mathbf{x}) := cl(\lambda\mathbf{x}), \qquad [1.27]$$

this linear space structure is what provides linearity to the canonical surjection.

PROOF.– It is a matter of demonstrating that the imposed definitions are independent from the chosen representatives. Thus, let $\mathbf{x}', \mathbf{y}'$ be two other representatives of $cl(\mathbf{x})$ and $cl(\mathbf{y})$. Since the relation $\mathtt{R}$ is compatible with addition, we have:

$$\mathbf{x}\mathtt{R}\mathbf{x}' \textit{ and } \mathbf{y}\mathtt{R}\mathbf{y}' \Rightarrow (\mathbf{x}+\mathbf{y})\,\mathtt{R}\,(\mathbf{x}'+\mathbf{y}') \Rightarrow cl(\mathbf{x}+\mathbf{y}) = cl(\mathbf{x}'+\mathbf{y}'). \qquad [1.28]$$

In the same way, we can write that:

$$\mathbf{x}\mathtt{R}\mathbf{x}' \Rightarrow \forall\lambda \in \mathbb{R},\ (\lambda\mathbf{x})\,\mathtt{R}\,(\lambda\mathbf{x}') \Rightarrow cl(\lambda\mathbf{x}) = cl(\lambda\mathbf{x}'). \qquad [1.29]$$

It is necessary to finish by demonstrating that addition and scalar multiplication thus defined satisfy the axioms of a linear space. It is left to the reader to verify these properties.

An interesting property of the canonical surjection:

PROPOSITION 1.9.– Let $\boldsymbol{E}/\boldsymbol{F}$ be the linear quotient space of $\boldsymbol{E}$ by (the linear subspace) $\boldsymbol{F}$. Then, we have:

$$cl(\mathbf{x}) = \mathbf{0} \Leftrightarrow \mathbf{x} \in \boldsymbol{F}. \qquad [1.30]$$

PROOF.– Suppose that $\mathbf{x} \in \boldsymbol{F}$. Then, for each $\mathbf{y} \in \boldsymbol{E}$, we have $(\mathbf{y}+\mathbf{x}) - \mathbf{y} \in \boldsymbol{F}$. We thus deduce that:

$$\forall \mathbf{x} \in \boldsymbol{F}, \forall \mathbf{y} \in \boldsymbol{E} \ \ (\mathbf{x}+\mathbf{y}) \mathrm{R} \mathbf{y} \Leftrightarrow \forall \mathbf{x} \in \boldsymbol{F}, \forall \mathbf{y} \in \boldsymbol{E}, \ cl(\mathbf{x}+\mathbf{y}) = cl(\mathbf{y}). \qquad [1.31]$$

Now, by definition of sum of a class we have $cl(\mathbf{x}+\mathbf{y}) = cl(\mathbf{x}) + cl(\mathbf{y})$, which entails:

$$\forall \mathbf{x} \in \boldsymbol{F}, \forall \mathbf{y} \in \boldsymbol{E}, \ \ cl(\mathbf{x}) + cl(\mathbf{y}) = cl(\mathbf{y}) \Leftrightarrow \forall \mathbf{x} \in \boldsymbol{F}, \ \ cl(\mathbf{x}) = \mathbf{0}. \qquad [1.32]$$

Inversely, let us suppose that we have $cl(\mathbf{x}) = \mathbf{0}$. Then, since it is clear that $cl(\mathbf{0}) = \mathbf{0}$, we hence deduce that $\mathbf{x}\mathrm{R}\mathbf{0}$, i.e. $\mathbf{x} - \mathbf{0} \in \boldsymbol{F}$ and thus $\mathbf{x} \in \boldsymbol{F}$.

1.4. Skew-symmetric multi-linear maps

1.4.1. *Permutations*

DEFINITION 1.8.– A p-index $\mathbf{i} := (i_1, \cdots, i_p)$ over the set I is a p-tuples of values, each chosen in I. Therefore, if $\mathbf{i}$ is a p-index, each of $i_k, k \in [1,p]$ has a value in I. We denote by $\mathcal{I}(p)$ the set of p-indexes.

In all that follows, we shall employ a special notation for the set of particular p-indexes:

DEFINITION 1.9.– When I is completely ordered, a p-index $\mathbf{i} \in \mathcal{I}(p)$ is said ordered (respectively strictly ordered) if:

$$i_1 \leq i_2 \leq \cdots \leq i_p \ \ (\textit{respectively} \ \ i_1 < i_2 < \cdots < i_p). \qquad [1.33]$$

We denote $\mathcal{I}_o(p) \subset \mathcal{I}(p)$ the set of p-indexes *strictly* ordered.

REMARK 1.3.– Let us note immediately the two following points:

1) If $I = [1, n]$ and if $p > n$, the set of strictly ordered p-indexes is empty: it is impossible to have a sequence of p values strictly ordered between 1 and n if each value is chosen in $[1, n]$.

2) If $I = [1, n]$ and if $p \leq n$, there exists exactly $\binom{n}{p}$ ways of constructing a strictly ordered p-index: for that purpose, it is sufficient to choose p elements among the set of values of $\{1, \cdots, n\}$. ■

Let us now consider the permutation notion.

DEFINITION 1.10.– Let $[1, n] := \{1, 2, \cdots, n\}$ be a set of integers from 1 to n. We call permutation over $[1, n]$ any bijective map of this set onto itself. For a fixed $n \geq 1$, we denote by S_n the set of all the permutations over $[1, n]$. There exists exactly $n!$ ways of constructing a permutation over $[1, n]$.

DEFINITION 1.11.– We call transposition over $[1, n]$ any permutation τ for which there exists $i \neq j \in [1, n]^2$ such that:

1) We have $\tau(i) = j$ and $\tau(j) = i$;

2) For all $k \in [1, n]$ such that $k \notin \{i, j\}$, we have $\tau(k) = k$.

Therefore, a transposition is a permutation that exchanges exactly two elements of $[1, n]$ while maintaining constant the others.

THEOREM 1.3.– [Decomposition of permutations by transpositions] Each permutation $\sigma \in S_n$ can be decomposed as a product (in the sense of composition) of transpositions. Such a decomposition is not unique, but the parity of the number of transpositions that decompose a permutation σ depends only on σ and not on the considered decomposition.

PROOF.– Conceded.

DEFINITION 1.12.– We define the signature $\varepsilon(\sigma)$ of a permutation $\sigma \in S_n$ as:

$$\varepsilon(\sigma) := (-1)^{N_t(\sigma)}, \qquad [1.34]$$

where $N_t(\sigma)$ is a number of transpositions that make it possible to decompose σ.

Before moving further, let us give immediately the essential property of the signature function:

THEOREM 1.4.– The signature map is a group morphism of $(S_n, \circ)$ into $(\{-1, 1\}, \times)$. In particular,

1) for all $\alpha, \beta \in S_n$, it satisfies $\varepsilon(\alpha \circ \beta) = \varepsilon(\alpha)\varepsilon(\beta)$;

2) for all $\alpha \in S_n$, we have $\varepsilon(\alpha^{-1}) = \varepsilon(\alpha)$.

PROOF.– This theorem is conceded (although there is no difficulty) once we concede the decomposition of permutations into a transposition product (see theorem 1.3). Let us note that the second statement is related to the fact that in $\{-1, 1\}$ each element is its own inverse with respect to multiplication.

1.4.2. *Alternating multi-linear map*

DEFINITION 1.13.– Let $\boldsymbol{E}$ be a linear space, $p \geq 2$ an integer and $\boldsymbol{F}$ a linear space. We say that a map $\psi_p : \boldsymbol{E}^p \mapsto \boldsymbol{F}$ is a:

1) p-linear map: if we fix $p-1$ variables in a list of arguments of ψ_p, then the (marginal) remaining map is a linear map:

$$\begin{aligned}&\psi_p(\mathbf{x}_1, \cdots, \mathbf{y}_k + \alpha\mathbf{z}_k, \cdots \mathbf{x}_p)\\&= \psi_p(\mathbf{x}_1, \cdots, \mathbf{y}_k, \cdots, \mathbf{x}_p) + \alpha\psi_p(\mathbf{x}_1, \cdots, \mathbf{z}_k, \cdots, \mathbf{x}_p).\end{aligned} \qquad [1.35]$$

2) Alternating: if, in a list of arguments of ψ_p, there are two equal vectors, then the image is zero:

$$\psi_p(\cdots, \mathbf{x}, \cdots, \mathbf{x}, \cdots) = \mathbf{0}. \qquad [1.36]$$

3) Skew-symmetric: if, in a list of arguments of ψ_p, we exchange two elements while maintaining the others constant, then the map becomes its own opposite:

$$\psi_p(\cdots, \mathbf{x}, \cdots, \mathbf{y}, \cdots) = -\psi_p(\cdots, \mathbf{y}, \cdots, \mathbf{x}, \cdots). \qquad [1.37]$$

PROPOSITION 1.10.– A p-linear map is skew-symmetric if and only if it is alternating.

PROOF.– Suppose that ψ_p is skew-symmetric. Imagine $\mathbf{x}_1, \cdots, \mathbf{x}_p$, a sequence of p arguments that contain two equal elements. We denote by $\mathbf{x}$ this element and we imagine that it is situated at position i and at position j. Then, it can be truly said that the list $\mathbf{x}_1, \cdots, \mathbf{x}_p$ is obtained starting from itself by exchanging the arguments that are situated at index i and at index j. By employing the fact that the map ψ_p is skew-symmetric we have thus:

$$\psi_p(\mathbf{x}_1, \cdots, \mathbf{x}_p) = -\psi_p(\mathbf{x}_1, \cdots, \mathbf{x}_p) \Leftrightarrow \psi_p(\mathbf{x}_1, \cdots, \mathbf{x}_p) = \mathbf{0}, \quad [1.38]$$

and thus the map is indeed alternating. Suppose now that the map is alternating. Then we can write that:

$$\psi_p(\cdots, \mathbf{x} + \mathbf{y}, \cdots, \mathbf{x} + \mathbf{y}, \cdots) = \mathbf{0}, \quad [1.39]$$

by developing this expression by considering the p-linearity, and by employing the fact that:

$$\psi_p(\cdots, \mathbf{x}, \cdots, \mathbf{x}, \cdots) = \psi_p(\cdots, \mathbf{y}, \cdots, \mathbf{y}, \cdots) = \mathbf{0}, \quad [1.40]$$

we then have:

$$\psi_p(\cdots, \mathbf{x}, \cdots, \mathbf{y}, \cdots) + \psi_p(\cdots, \mathbf{y}, \cdots, \mathbf{x}, \cdots) = \mathbf{0}, \quad [1.41]$$

which shows that the map ψ_p is skew-symmetric.

In practice, there is no inconvenience in exchanging the terms alternating and skew-symmetric in order to demonstrate one a property for the latter. We shall often employ the substitution of both terms in practice.

PROPOSITION 1.11.– Let $\sigma \in S_p$ be a permutation over $[1, p]$, ψ_p a p-linear alternating map of $\boldsymbol{E}^p$ in a space $\boldsymbol{F}$ and finally $\mathbf{x}_1, \cdots, \mathbf{x}_p$ a sequence of p vectors of $\boldsymbol{E}$. Then, we have:

$$\psi_p\left(\mathbf{x}_{\sigma(1)}, \cdots, \mathbf{x}_{\sigma(p)}\right) = \varepsilon(\sigma)\, \psi_p(\mathbf{x}_1, \cdots, \mathbf{x}_p). \quad [1.42]$$

PROOF.– Immediate. The decomposition of σ into a transposition product $\tau_1 \circ \cdots \circ \tau_{N_t(\sigma)}$ is done by using theorem 1.3. We have then:

$$\psi_p\left(\mathbf{x}_{\sigma(1)}, \cdots, \mathbf{x}_{\sigma(p)}\right) = \psi_p\left(\mathbf{x}_{\tau_1 \circ \cdots \tau_{N_t}(\sigma)(1)}, \cdots, \mathbf{x}_{\tau_1 \circ \cdots \tau_{N_t}(\sigma)(p)}\right), \quad [1.43]$$

then we write:

$$\psi_p\left(\mathbf{x}_{\tau_1\circ\cdots\tau_{N_t}(\sigma)(1)},\cdots,\mathbf{x}_{\tau_1\circ\cdots\tau_{N_t}(\sigma)(p)}\right) = -\psi_p\left(\mathbf{x}_{\tau_2\circ\cdots\tau_{N_t}(\sigma)(1)},\cdots,\mathbf{x}_{\tau_2\circ\cdots\tau_{N_t}(\sigma)(p)}\right). \quad [1.44]$$

By repeating the calculation we quickly obtain:

$$\psi_p\left(\mathbf{x}_{\sigma(1)},\cdots,\mathbf{x}_{\sigma(p)}\right) = (-1)^{N_t(\sigma)}\,\psi_p\left(\mathbf{x}_1,\cdots,\mathbf{x}_p\right) = \varepsilon(\sigma)\,\psi_p\left(\mathbf{x}_1,\cdots,\mathbf{x}_p\right). \quad [1.45]$$

Let us finish with a short proposition regarding the multi-linear aspect:

PROPOSITION 1.12.– Let $\psi : \boldsymbol{E}^p \mapsto \boldsymbol{F}$ be a map. Then, it is p-linear if and only if for any sequence of vectors $\mathbf{x}_{i_k}^k, k \in [1,p], i_k \in [1,p]$ and for any sequence of scalars $\alpha_{i_k}^k, k \in [1,p], i_k \in [1,p]$ we have:

$$\psi_p\left(\sum_{i_1}\alpha_{i_1}^1\mathbf{x}_{i_1}^1,\cdots,\sum_{i_p}\alpha_{i_p}^p\mathbf{x}_{i_p}^p\right) = \sum_{i_1,\cdots,i_p}\alpha_{i_1}^1,\ldots\alpha_{i_p}^p\,\psi_p\left(\mathbf{x}_{i_1}^1,\cdots,\mathbf{x}_{i_p}^p\right). \quad [1.46]$$

PROOF.– Immediate.

1.4.3. *Identification of bilinear maps*

PROPOSITION 1.13.– Let $\boldsymbol{E},\boldsymbol{F},\boldsymbol{G}$ be three linear spaces. We denote by $\mathcal{M}_2(\boldsymbol{E}\times\boldsymbol{F},\boldsymbol{G})$ the space of bilinear maps of $\boldsymbol{E}\times\boldsymbol{F}$ with values in $\boldsymbol{G}$, then:

$\mathcal{M}_2(\boldsymbol{E}\times\boldsymbol{F},\boldsymbol{G})$ *is isomorphic to* $\mathcal{L}(\boldsymbol{E},\mathcal{L}(\boldsymbol{F},\boldsymbol{G}))$. [1.47]

PROOF.– Let $B \in \mathcal{M}_2(\boldsymbol{E}\times\boldsymbol{F},\boldsymbol{G})$. Then, for a fixed $\mathbf{x}$ in $\boldsymbol{E}$, the map $B(\mathbf{x},\cdot)$ is indeed a linear map of $\boldsymbol{F}$ into $\boldsymbol{G}$. Furthermore, the dependence of $B(\mathbf{x},\cdot)$ with $\mathbf{x}$ is indeed linear, which makes it possible to actually define in a unique way, starting from B, a map b which is linear, from $\boldsymbol{E}$ with values in $\mathcal{L}(\boldsymbol{F},\boldsymbol{G})$. We denote by ψ this map. Such a map $\psi : B \mapsto b$ is trivially linear. Furthermore, it is invertible. In fact, suppose that we have $b \in \mathcal{L}(\boldsymbol{E},\mathcal{L}(\boldsymbol{F},\boldsymbol{G}))$. Consider then:

$$B : \mathbf{x},\mathbf{y} \mapsto b[\mathbf{x}](\mathbf{y}), \quad [1.48]$$

it is very clear that $B(\cdot,\cdot)$ is a bilinear map over $\boldsymbol{E} \times \boldsymbol{F}$ with values in $\boldsymbol{G}$. On the other hand, by very construction of B we have $\psi(B) = b$. The map ψ is thus surjective. Now, if we have $\psi(B) = 0[\cdot](\cdot)$, this means that:

$$\forall \mathbf{x} \in \boldsymbol{E}, \psi(B)[\mathbf{x}](\cdot) = 0(\cdot) \in \mathcal{L}(\boldsymbol{F}, \boldsymbol{G}). \quad [1.49]$$

We have seen before that we had:

$$B(\mathbf{x}, \mathbf{y}) = \psi(B)[\mathbf{x}](\mathbf{y}). \quad [1.50]$$

Thus, if $\psi(B)$ is zero, this means that:

$$\forall \mathbf{x}, \mathbf{y}, \ \psi(B)[\mathbf{x}](\mathbf{y}) = 0(\mathbf{y}) = \mathbf{0}, \quad [1.51]$$

and thus B is indeed a null bilinear map, which shows that ψ is injective.

DEFINITION 1.14.– Let $\boldsymbol{E}_1, \cdots, \boldsymbol{E}_p, \boldsymbol{F}$ be some linear spaces. We denote by:

$$\mathcal{M}_p(\boldsymbol{E}_1 \times \cdots \times \boldsymbol{E}_p, \boldsymbol{F})$$

the (linear) space of p-linear maps over $\boldsymbol{E}_1 \times \cdots \times \boldsymbol{E}_p$ with values in $\boldsymbol{F}$.

2

Construction of Exterior Algebras

In this chapter, we describe the existence of exterior algebras as a structure that enables the resolution of a universal problem, which can be expressed in the following way:

QUESTION 2.1.– Let $\boldsymbol{E}$ be a linear space. Does there exist a set which enables the factorization of any multi-linear skew-symmetric map over $\boldsymbol{E}$, by means of linear maps?

A positive answer to this question allows us to formulate the following statement, which essentially summarizes the philosophy of exterior algebras: *any p-linear skew-symmetric map over vectors can necessarily and uniquely be written as a linear map over some p-vectors.* Therefore, p-vectors will be the new objects that we shall construct in this chapter: they are precisely the elements of exterior algebras. Thus, a specific exterior algebra will be associated with each degree of multi-linearity, with a degree corresponding to that of the multi-linear maps factorized with this method.

REMARK 2.1.– It is important to be aware that such new objects (p-vectors) have absolutely no simple equivalent object. They represent completely new entities. In particular, they definitely cannot be reduced to simple Cartesian products of vectors. ■

2.1. Existence

2.1.1. *Construction theorem statement*

THEOREM 2.1 (existence of exterior algebra).– Let $\boldsymbol{E}$ be a linear space and $p \geq 2$ an integer. Then, there exists, respectively,

1) a linear space $\boldsymbol{F}$ and

2) a skew-symmetric p-linear map $\psi_p : \boldsymbol{E}^P \mapsto \boldsymbol{F}$,

such that

1) if $\boldsymbol{G}$ is another linear space,

2) $\gamma_p : \boldsymbol{E}^P \mapsto \boldsymbol{G}$ is another skew-symmetric p-linear map,

and there exists a unique linear map $l : \boldsymbol{F} \mapsto \boldsymbol{G}$ such that we have $\gamma_p = l \circ \psi_p$.

This existence theorem is usually summarized by means of the following factorization diagram, where ψ_p and γ_p are alternating p-linear maps, l is a linear map and $\boldsymbol{F}$ is a set of p-vectors:

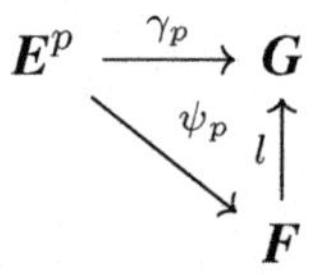

where we finally have $\gamma_p = l \circ \psi_p$, with the relation between l and γ_p being dependent on the construction operator ψ_p.

2.1.2. *Demonstration for* $p = 2$

PROOF.– At first, we develop the demonstration for $p = 2$, and we shall show later how it is generalized for any p. Let us call $\mathcal{F}_0\left(\boldsymbol{E} \times \boldsymbol{E} \mapsto \mathbb{R}\right)$ the set of maps over $\boldsymbol{E} \times \boldsymbol{E}$ with values in $\mathbb{R}$ that are zero everywhere except in a finite number of points. We define the map $\lambda : \boldsymbol{E} \times \boldsymbol{E} \mapsto \mathcal{F}_0\left(\boldsymbol{E} \times \boldsymbol{E} \mapsto \mathbb{R}\right)$ such that for $\mathbf{x}, \mathbf{y}$ in $\boldsymbol{E}$:

$$\lambda_{\mathbf{x},\mathbf{y}}\left(\mathbf{r},\mathbf{s}\right) := 1 \ \textit{if} \ \left(\mathbf{r},\mathbf{s}\right) = \left(\mathbf{x},\mathbf{y}\right), \ \ \lambda_{\mathbf{x},\mathbf{y}}\left(\mathbf{r},\mathbf{s}\right) := 0 \ \textit{if} \ \left(\mathbf{r},\mathbf{s}\right) \neq \left(\mathbf{x},\mathbf{y}\right). \qquad [2.1]$$

Note that, in accordance with the considerations of proposition 1.6, the family of $\lambda_{\mathbf{x},\mathbf{y}}(\cdot,\cdot)$ is an algebraic basis of the linear space $\mathcal{F}_0(\boldsymbol{E}\times\boldsymbol{E}\mapsto\mathbb{R})$. Within this space, we define $\mathcal{G}$ as the set of all functions of the following form:

$$\lambda_{\alpha\mathbf{x}+\beta\mathbf{y},\gamma\mathbf{z}+\delta\mathbf{t}} + \frac{1}{2}\left(\alpha\gamma\lambda_{\mathbf{z},\mathbf{x}} + \alpha\delta\lambda_{\mathbf{t},\mathbf{x}} + \beta\gamma\lambda_{\mathbf{z},\mathbf{y}} + \beta\delta\lambda_{\mathbf{t},\mathbf{y}}\right) - \frac{1}{2}\left(\alpha\gamma\lambda_{\mathbf{x},\mathbf{z}} + \alpha\delta\lambda_{\mathbf{x},\mathbf{t}} + \beta\gamma\lambda_{\mathbf{y},\mathbf{z}} + \beta\delta\lambda_{\mathbf{y},\mathbf{t}}\right), \quad [2.2]$$

when $\mathbf{x},\mathbf{y},\mathbf{z},\mathbf{t}$ span all the quadruplets of $\boldsymbol{E}^4$ and $\alpha,\beta,\gamma,\delta$ span the set of all the quadruplets of $\mathbb{R}^4$. We denote by $\langle\mathcal{G}\rangle$ the linear subspace generated by $\mathcal{G}$ within the space $\mathcal{F}_0(\boldsymbol{E}\times\boldsymbol{E}\mapsto\mathbb{R})$. Within the latter linear space, we define the relation R with the expression:

$$\forall f,g\in\mathcal{F}_0(\boldsymbol{E}\times\boldsymbol{E}\mapsto\mathbb{R}),\ f\mathrm{R}g \Leftrightarrow f-g\in\langle\mathcal{G}\rangle. \quad [2.3]$$

Since $\langle\mathcal{G}\rangle$ is a linear subspace of $\mathcal{F}_0(\boldsymbol{E}\times\boldsymbol{E}\mapsto\mathbb{R})$, we know that the relation R is an equivalence relation (see example 1.1), compatible with pointwise addition and scalar multiplication in $\mathcal{F}_0(\boldsymbol{E}\times\boldsymbol{E}\mapsto\mathbb{R})$ (see example 1.2), that is:

$$\forall\lambda\in\mathbb{R},\ \forall f,g\in\mathcal{F}_0(\boldsymbol{E}\times\boldsymbol{E}\mapsto\mathbb{R}),\ f\mathrm{R}g\Rightarrow(\lambda f)\,\mathrm{R}\,(\lambda g) \quad [2.4]$$

$$\forall f,g,f',g'\in\mathcal{F}_0(\boldsymbol{E}\times\boldsymbol{E}\mapsto\mathbb{R}),\ f\mathrm{R}g \text{ and } f'\mathrm{R}g' \Rightarrow (f+f')\,\mathrm{R}\,(g+g') \quad [2.5]$$

As a result, the quotient space $\mathcal{F}_0(\boldsymbol{E}\times\boldsymbol{E}\mapsto\mathbb{R})/\mathrm{R}$ can be equipped with the structure of a linear space (see theorem 1.2). By definition, we denote by $\boldsymbol{F}$ this space (which will be the space of the theorem) and we denote by J the canonical surjection of $\mathcal{F}(\boldsymbol{E}\times\boldsymbol{E}\mapsto\mathbb{R})$ in its quotient space $\boldsymbol{F}$, which, as we recall, admits exactly $\langle\mathcal{G}\rangle$ as kernel (see proposition 1.9). Then, the map $\psi_2 : J\circ\lambda : \boldsymbol{E}\times\boldsymbol{E}\mapsto\boldsymbol{F}$ is:

1) An alternating map. In fact, let us calculate:

$$J(\lambda_{\mathbf{x},\mathbf{y}}) + J\left(\frac{1}{2}\lambda_{\mathbf{y},\mathbf{x}}\right) - J\left(\frac{1}{2}\lambda_{\mathbf{x},\mathbf{y}}\right). \quad [2.6]$$

Due to the linearity of J, it is equal to:

$$J\left(\lambda_{\mathbf{x},\mathbf{y}} + \frac{1}{2}\lambda_{\mathbf{y},\mathbf{x}} - \frac{1}{2}\lambda_{\mathbf{x},\mathbf{y}}\right) = J\left(\frac{1}{2}\lambda_{\mathbf{x},\mathbf{y}} + \frac{1}{2}\lambda_{\mathbf{y},\mathbf{x}}\right). \quad [2.7]$$

However, the left-hand side of this equality is null, because of the fact that J is linear with kernel $\langle\mathcal{G}\rangle$ and the function:

$$\lambda_{\mathbf{x},\mathbf{y}} + \frac{1}{2}\lambda_{\mathbf{y},\mathbf{x}} - \frac{1}{2}\lambda_{\mathbf{x},\mathbf{y}} \quad [2.8]$$

belongs to $\langle \mathcal{G} \rangle$ since it has the form of [2.2] with

$$(\mathbf{x}, \mathbf{y}, \mathbf{z}, \mathbf{t}) \equiv (\mathbf{x}, \mathbf{0}, \mathbf{y}, \mathbf{0}), (\alpha, \beta, \gamma, \delta) \equiv (1, 0, 1, 0). \quad [2.9]$$

We deduce thereof that the right-hand side is null and that the function ψ_2 is skew-symmetric:

$$J(\lambda_{\mathbf{x},\mathbf{y}} + \lambda_{\mathbf{y},\mathbf{x}}) = 0 \Leftrightarrow \psi_2(\mathbf{x}, \mathbf{y}) = -\psi_2(\mathbf{y}, \mathbf{x}). \quad [2.10]$$

2) A 2-linear map. In order to verify this bilinearity, it actually is enough to show the bilinearity of the left-hand side, in virtue of the skew symmetry. In the same way as in the previous demonstration, we calculate:

$$J\left(\lambda_{\mathbf{x}+\alpha\mathbf{y},\mathbf{z}} + \frac{1}{2}[\lambda_{\mathbf{z},\mathbf{x}} + \alpha\lambda_{\mathbf{z},\mathbf{y}}] - \frac{1}{2}[\lambda_{\mathbf{x},\mathbf{z}} + \alpha\lambda_{\mathbf{x},\mathbf{z}}]\right) = \mathbf{0}, \quad [2.11]$$

since the argument function of J has the form of [2.2] with

$$(\mathbf{x}, \mathbf{y}, \mathbf{z}, \mathbf{t}) \equiv (\mathbf{x}, \mathbf{y}, \mathbf{z}, \mathbf{t}), (\alpha, \beta, \gamma, \delta) \equiv (1, \alpha, 1, 0), \quad [2.12]$$

we immediately obtain:

$$\psi_2(\mathbf{x} + \alpha\mathbf{y}, \mathbf{z}) + \psi_2(\mathbf{z}, \mathbf{x}) + \alpha\psi_2(\mathbf{z}, \mathbf{y}) = \mathbf{0} \Leftrightarrow$$
$$\psi_2(\mathbf{x} + \alpha\mathbf{y}, \mathbf{z}) = -\psi_2(\mathbf{z}, \mathbf{x}) - \alpha\psi_2(\mathbf{z}, \mathbf{y}), \quad [2.13]$$

and we conclude then by employing the skew symmetry of ψ_2 in the right-hand side.

We shall now show that the space $\boldsymbol{F}$ and the map ψ_2 satisfy definition 2.1. Thus, let $\boldsymbol{G}$ be a linear space and γ_2 be a skew-symmetric map of $\boldsymbol{E} \times \boldsymbol{E}$ into $\boldsymbol{G}$. Let us consider within $\mathcal{F}_0(\boldsymbol{E} \times \boldsymbol{E} \mapsto \mathbb{R})$ the algebraic basis $\lambda_{\mathbf{x},\mathbf{y}}, \mathbf{x} \in \boldsymbol{E}, \mathbf{y} \in \boldsymbol{E}$ and consider in $\boldsymbol{G}$ the family $\gamma_2(\mathbf{x}, \mathbf{y}), \mathbf{x} \in \boldsymbol{E}, \mathbf{y} \in \boldsymbol{E}$. Then, according to proposition 1.4, there exists a unique linear map $L : \mathcal{F}_0(E \times E \mapsto \mathbb{R}) \mapsto \boldsymbol{G}$ such that we have:

$$\forall \mathbf{x}, \mathbf{y} \in \boldsymbol{E}, \quad L(\lambda_{\mathbf{x},\mathbf{y}}) = \gamma_2(\mathbf{x}, \mathbf{y}). \quad [2.14]$$

On the other hand, it is clear that if $f, g \in \mathcal{F}_0(\boldsymbol{E} \times \boldsymbol{E} \mapsto \mathbb{R})$ are in relation for $\langle \mathcal{G} \rangle$ (in other words, we have $f\mathrm{R}g$), then we also have $L(f) = L(g)$. In order to verify that, it suffices to observe that any function of $\langle \mathcal{G} \rangle$ is null for L. Now, we have:

$$L\Big(\sum\Big[\lambda_{\alpha\mathbf{x}+\beta\mathbf{y},\gamma\mathbf{z}+\delta\mathbf{t}} + \frac{1}{2}(\alpha\gamma\lambda_{\mathbf{z},\mathbf{x}} + \alpha\delta\lambda_{\mathbf{t},\mathbf{x}} + \beta\gamma\lambda_{\mathbf{z},\mathbf{y}} + \beta\delta\lambda_{\mathbf{t},\mathbf{y}})$$
$$- \frac{1}{2}(\alpha\gamma\lambda_{\mathbf{x},\mathbf{z}} + \alpha\delta\lambda_{\mathbf{x},\mathbf{t}} + \beta\gamma\lambda_{\mathbf{y},\mathbf{z}} + \beta\delta\lambda_{\mathbf{y},\mathbf{t}})\Big]\Big) \quad [2.15]$$

which is equal to

$$\sum \gamma_2 (\alpha \mathbf{x} + \beta \mathbf{y}, \gamma \mathbf{z} + \delta \mathbf{t})$$
$$+ \frac{1}{2} [\alpha\gamma\gamma_2 (\mathbf{z}, \mathbf{x}) + \alpha\delta\gamma_2 (\mathbf{t}, \mathbf{x}) + \beta\gamma\gamma_2 (\mathbf{z}, \mathbf{y}) + \beta\delta\gamma_2 (\mathbf{t}, \mathbf{y})]$$
$$- \frac{1}{2} [\alpha\gamma\gamma_2 (\mathbf{x}, \mathbf{z}) + \alpha\delta\gamma_2 (\mathbf{x}, \mathbf{t}) + \beta\gamma\gamma_2 (\mathbf{y}, \mathbf{z}) + \beta\delta\gamma_2 (\mathbf{y}, \mathbf{t})] \quad [2.16]$$

which is also null in virtue of the bilinearity and skew symmetry of function γ_2. Therefore, starting from L we can define a unique map $l : \boldsymbol{F} \mapsto \boldsymbol{G}$ such that:

$$\forall \mathbf{X} \in \boldsymbol{F}, l(\mathbf{X}) = L(\lambda_{\mathbf{x},\mathbf{y}}), \quad [2.17]$$

where $\lambda_{\mathbf{x},\mathbf{y}}$ is a representative of $\mathbf{X}$ in $\mathcal{F}_0 (\boldsymbol{E} \times \boldsymbol{E} \mapsto \mathbb{R})$. We then have:

$$\forall \mathbf{x}, \mathbf{y} \in \boldsymbol{E}, \quad l \circ (J(\lambda_{\mathbf{x},\mathbf{y}})) = \gamma_2 (\mathbf{x}, \mathbf{y}), \quad [2.18]$$

that is, we indeed have $l \circ \psi_2 = \gamma_2$. We must now verify that the map l is unique. In fact, we know that the family $J(\lambda_{\mathbf{x},\mathbf{y}}), \mathbf{x} \in \boldsymbol{E}, \mathbf{y} \in \boldsymbol{E}$ is a generating set of $\boldsymbol{F}$ since J is surjective from $\mathcal{F}_0 (\boldsymbol{E} \times \boldsymbol{E} \mapsto \mathbb{R})$ into $\boldsymbol{F}$ and since $\{\lambda_{\mathbf{x},\mathbf{y}}, \mathbf{x} \in \boldsymbol{E}, \mathbf{y} \in \boldsymbol{E}\}$ is an algebraic basis of $\mathcal{F}_0 (\boldsymbol{E} \times \boldsymbol{E} \mapsto \mathbb{R})$ (see proposition 1.2). According to the incomplete basis theorem (see corollary 1.1), we can thus extract from this family an algebraic basis $J(\lambda_{\mathbf{x}_i,\mathbf{y}_i}), i \in I$ of $\boldsymbol{F}$. Then, we have:

$$\forall \mathbf{x} \in \boldsymbol{E}, \forall \mathbf{y} \in \boldsymbol{E}, \quad l(\psi_2 (\mathbf{x}, \mathbf{y})) = \gamma_2 (\mathbf{x}, \mathbf{y}) \Rightarrow$$
$$\forall i \in I, \quad l(\psi_2 (\mathbf{x}_i, \mathbf{y}_i)) = \gamma_2 (\mathbf{x}_i, \mathbf{y}_i). \quad [2.19]$$

In particular, thanks to proposition 1.4, we know that there exists a unique application $l : \boldsymbol{F} \mapsto \boldsymbol{G}$ that can satisfy this property. Hence the sought after uniqueness of the map.

2.1.3. *Demonstration for* $p > 2$

QUESTION 2.2.– How can this demonstration be generalized for $p > 2$?

1) Let us consider the set $\mathcal{F}_0 (\boldsymbol{E}^p \mapsto \mathbb{R})$ of functions of $\boldsymbol{E}^p$ with values in $\mathbb{R}$ which are zero everywhere except in a finite number of points, equipped with its algebraic basis (see proposition 1.6) $\lambda_{\mathbf{x}_1,\cdots,\mathbf{x}_p} (\cdot, \cdots, \cdot)$, $(\mathbf{x}_1, \cdots, \mathbf{x}_p) \in \boldsymbol{E}^p$:

$$\forall \mathbf{y}_1, \cdots, \mathbf{y}_p, \quad \lambda_{\mathbf{x}_1,\cdots,\mathbf{x}_p} (\mathbf{y}_1, \cdots, \mathbf{y}_p) = \begin{cases} 1 & \textit{if } \forall i \ \mathbf{x}_i = \mathbf{y}_i, \\ 0 & \textit{if } \exists i \ \mathbf{x}_i \neq \mathbf{y}_i \end{cases}. \quad [2.20]$$

2) Then, we consider in $\mathcal{F}_0\left(\boldsymbol{E}^p \mapsto \mathbb{R}\right)$ the set $\mathcal{G}$ of maps of the form:

$$\left(\lambda_{\Sigma_{i_1} \alpha^1_{i_1} \mathbf{x}^1_{i_1}, \cdots, \Sigma_{i_p} \alpha^p_{i_p} \mathbf{x}^p_{i_p}}\right) - \frac{1}{p!} \sum_{i_1, \cdots, i_p} \alpha^1_{i_1} \cdots \alpha^p_{i_p} \sum_{\sigma \in S_p} \epsilon(\sigma)\, \lambda_{\sigma\left(\mathbf{x}^1_{i_1}, \cdots, \mathbf{x}^p_{i_p}\right)}, \quad [2.21]$$

where $\sigma\left(\mathbf{y}_1, \cdots, \mathbf{y}_p\right) = \left(\mathbf{y}_{\sigma(1)}, \cdots, \mathbf{y}_{\sigma(p)}\right)$ where each $\mathbf{x}^k_{i_k}$ is in $\boldsymbol{E}$, each $\alpha^k_{i_k}$ is in $\mathbb{R}$, the indexes $i_k, k \in [1, p]$ span each from 1 to p and the symbol $\sigma \in S_p$ represents the fact that we span all the permutations of $[1, p]$. We denote by $\langle \mathcal{G} \rangle$ the linear subspace of $\mathcal{F}_0\left(\boldsymbol{E}^p \mapsto \mathbb{R}\right)$ generated by the set $\mathcal{G}$.

3) We set $\boldsymbol{F}$ to be the quotient space of $\mathcal{F}_0\left(\boldsymbol{E}^p \mapsto \mathbb{R}\right)$ quotiented by $\langle \mathcal{G} \rangle$ and we consider the canonical surjection J of $\mathcal{F}_0\left(\boldsymbol{E}^p \mapsto \mathbb{R}\right)$ in $\boldsymbol{F}$.

4) We show then that the map $J\left(\lambda_{\mathbf{x}_1, \cdots, \mathbf{x}_p}\right)$ satisfies:

$$\forall \pi \in S_p, \quad J\left(\lambda_{\mathbf{x}_{\pi(1)}, \cdots, \mathbf{x}_{\pi(p)}}\right) = \epsilon(\pi)\, J\left(\lambda_{\mathbf{x}_1, \cdots, \mathbf{x}_p}\right). \quad [2.22]$$

We denote by definition $\psi_p\left(\mathbf{x}_1, \cdots, \mathbf{x}_p\right) := J\left(\lambda_{\mathbf{x}_1, \cdots, \mathbf{x}_p}\right)$ which is therefore a skew-symmetric map defined over $\boldsymbol{E}^p$ with values in $\boldsymbol{F}$.

5) We then show that the map ψ_p of $\boldsymbol{E}^p$ with values in $\boldsymbol{F}$ is p-linear.

6) Afterwards, we consider a map $\gamma_p : \boldsymbol{E}^p \mapsto \boldsymbol{G}$ that is skew-symmetric. Following the same arguments as for the case $p = 2$, there exists a unique linear map L that maps $\lambda_{\mathbf{x}_1, \cdots, \mathbf{x}_p}$ onto $\gamma_p\left(\mathbf{x}_1, \cdots, \mathbf{x}_p\right)$.

7) An explicit calculation shows that this linear map L is equal to zero over the linear space $\langle \mathcal{G} \rangle$.

8) By means of the same arguments as in the case $p = 2$, thanks to a quotient operation this linear map enables the definition of a linear map $l : \boldsymbol{F} \mapsto \boldsymbol{G}$ such that $l\left(\psi_p\left(\mathbf{x}_1, \cdots, \mathbf{x}_p\right)\right) = \gamma_p\left(\mathbf{x}_1, \cdots, \mathbf{x}_p\right)$.

9) The same arguments as in the case $p = 2$ show that the existence of a linear map that satisfies the previous equality is unique.

PROOF.– Actually, only the points [4, 5] might not be immediate compared to the demonstration regarding $p = 2$, since all the other points can be proved in a similar way. In both cases, the demonstration of statements [4, 5] consists essentially of a task of writing the formulas (as often happens when symmetry properties have to be shown). Therefore, it is not completely useless to dedicate some time to perform this writing work in order to make the argumentation very clear.

NOTATIONS.– Let us start with some notations: let M be "a square matrix", i.e. a twice indexed sequence $M_{i,j}$ with $i \in [1, p]$ and $j \in [1, p]$. For any permutation π over

$[1,p]$ (i.e. $\pi \in S_p$), we define the action of π on M, denoted by $\pi(M)$ according to the expression:

$$[\pi(M)]_{i,j} = M_{\pi(i),j}. \quad [2.23]$$

In other words, the effect of π is a permutation of rows of the matrix. The families $\alpha^k_{i_k}, k \in [1,p], i_k \in [1,p]$ as well as $\mathbf{x}^k_{i_k}, k \in [1,p], i_k \in [1,p]$ can be placed into matrices $A, \mathbf{X}$ defined in the following way:

$$\forall i \in [1,p],\ j \in [1,p],\ \ A_{i,j} := \alpha^i_j,\ \ \mathbf{X}_{i,j} := \mathbf{x}^i_j. \quad [2.24]$$

Finally, we employ the following notations:

$$\lambda_{\sum_{i_1} \alpha^1_{i_1} \mathbf{x}^1_{i_1}, \cdots, \sum_{i_p} \alpha^p_{i_p} \mathbf{x}^p_{i_p}} = \lambda[A, \mathbf{X}] \quad [2.25]$$

$$\frac{1}{p!} \sum_{i_1,\cdots,i_p} \alpha^1_{i_1} \cdots \alpha^p_{i_p} \sum_{\sigma \in S_p} \varepsilon(\sigma) \lambda_{\sigma\left(\mathbf{x}^1_{i_1},\cdots,\mathbf{x}^p_{i_p}\right)} := \beta[A, \mathbf{X}] \quad [2.26]$$

Justification of skew symmetry. Note that the very construction of the set $\mathcal{G}$ guarantees that for any pair of matrices $A, \mathbf{X}$ the difference $\lambda[A,\mathbf{X}] - \beta[A,\mathbf{X}]$ is an element of $\mathcal{G}$ and thus of $\langle \mathcal{G} \rangle$. In particular, using the notations of relation for the linear subspace $\langle \mathcal{G} \rangle$ we have:

$$\forall A,\ \forall \mathbf{X},\ \ \lambda[A,\mathbf{X}] \,\mathrm{R}\, \beta[A,\mathbf{X}]. \quad [2.27]$$

We have then the following lemma:

LEMMA 2.1.– For any pair $A, \mathbf{X}$ of squared matrices of reals and of vectors, for any permutation $\pi \in S_p$, we have $\beta[A, \pi(\mathbf{X})] = \varepsilon(\pi) \beta[A, \mathbf{X}]$, where $\varepsilon(\pi)$ is the signature of the permutation π.

PROOF.– Let us write:

$$\beta[A, \pi(\mathbf{X})] = \frac{1}{p!} \sum_{i_1,\cdots,i_p} \alpha^1_{i_1} \cdots \alpha^p_{i_p} \sum_{\sigma \in S_p} \varepsilon(\sigma) \lambda_{\sigma\left(\pi\left(\mathbf{x}^1_{i_1},\cdots,\mathbf{x}^p_{i_p}\right)\right)}. \quad [2.28]$$

By means of a change of variable $\sigma \circ \pi := \varsigma$ in the summation over the permutations, we immediately have:

$$\beta[A, \pi(\mathbf{X})] = \frac{1}{p!} \sum_{i_1,\cdots,i_p} \alpha^1_{i_1} \cdots \alpha^p_{i_p} \sum_{\varsigma \in S_p} \varepsilon\left(\varsigma \circ \pi^{-1}\right) \lambda_{\varsigma\left(\mathbf{x}^1_{i_1},\cdots,\mathbf{x}^p_{i_p}\right)}. \quad [2.29]$$

Now, we know that for any permutation $\pi, \varsigma \in S_p$ we have the following equalities (see theorem 1.4):

$$\epsilon\left(\varsigma \circ \pi^{-1}\right) = \varepsilon\left(\pi^{-1}\right)\epsilon\left(\pi\right) = \varepsilon\left(\varsigma\right)\epsilon\left(\pi\right). \qquad [2.30]$$

We can deduce that:

$$\begin{aligned} \beta\left[A, \pi\left(\mathbf{X}\right)\right] &= \varepsilon\left(\pi\right)\frac{1}{p!}\sum_{i_1,\cdots,i_p} \alpha_{i_1}^1 \cdots \alpha_{i_p}^p \sum_{\varsigma \in S_p} \epsilon\left(\zeta\right)\lambda_{\zeta\left(\mathbf{x}_{i_1}^1,\cdots,\mathbf{x}_{i_p}^p\right)} \\ &= \varepsilon\left(\pi\right)\beta\left(A, \mathbf{X}\right). \end{aligned} \qquad [2.31]$$

At this point, it is possible to complete the proof of skew symmetry, and for this purpose we begin with the fact that:

$$\forall A, \forall \mathbf{X}, \forall \pi \in S_p, \ \ \lambda\left[A, \mathbf{X}\right] \mathrm{R} \beta\left[A, \mathbf{X}\right] \ \ \textit{and} \ \lambda\left[A, \pi\left(\mathbf{X}\right)\right] \mathrm{R} \beta\left[A, \pi\left(\mathbf{X}\right)\right]. \qquad [2.32]$$

Since the relation R is compatible with the scalar multiplication (indeed, it is an equivalence relation constructed over a linear subspace), we deduce from the first relation, thanks to a multiplication by the signature $\varepsilon\left(\pi\right)$, that:

$$\lambda\left[A, \mathbf{X}\right] \mathrm{R} \beta\left[A, \mathbf{X}\right] \Rightarrow \epsilon\left(\pi\right)\lambda\left[A, \mathbf{X}\right] \mathrm{R} \epsilon\left(\pi\right)\beta\left[A, \mathbf{X}\right]. \qquad [2.33]$$

Then, by employing the equality [2.31], we deduce from the second relation:

$$\lambda\left[A, \pi\left(\mathbf{X}\right)\right] \mathrm{R} \beta\left[A, \pi\left(\mathbf{X}\right)\right] \Rightarrow \lambda\left[A, \pi\left(\mathbf{X}\right)\right] \mathrm{R} \varepsilon\left(\pi\right)\beta\left[A, \mathbf{X}\right]. \qquad [2.34]$$

Using then the transitivity of the relation R, we deduce from the two last relations that we indeed have:

$$\varepsilon\left(\pi\right)\lambda\left[A, \mathbf{X}\right] \mathrm{R} \lambda\left[A, \pi\left(\mathbf{X}\right)\right]. \qquad [2.35]$$

Now, we choose two particular matrices as A and $\mathbf{X}$:

$$\forall i \in [1, n], \ \begin{cases} A_{i,1} = 1 & \mathbf{X}_{i,1} = \mathbf{x}_i \\ A_{i,j} = 0 \ \ j > 1 & \mathbf{X}_{i,j} = \mathbf{0} \ \ j > 1 \end{cases}, \qquad [2.36]$$

which entails

$$\forall i \in [1, n], \ \begin{cases} A_{i,1} = 1 & \pi\left(\mathbf{X}\right)_{i,1} = \mathbf{x}_{\pi(i)} \\ A_{i,j} = 0 \ \ j > 1 & \pi\left(\mathbf{X}\right)_{i,j} = \mathbf{0} \ \ j > 1 \end{cases}. \qquad [2.37]$$

We apply the result [2.35] and we obtain that:

$$\lambda_{\pi(\mathbf{x}_1,\cdots,\mathbf{x}_p)} \mathrm{R} \varepsilon(\pi) \lambda_{\mathbf{x}_1,\cdots,\mathbf{x}_p}. \qquad [2.38]$$

By definition, this implies, by applying the quotient and since the canonical surjection is linear, that:

$$J\left(\lambda_{\pi(\mathbf{x}_1,\cdots,\mathbf{x}_p)}\right) = \varepsilon(\pi) J\left(\lambda_{\mathbf{x}_1,\cdots,\mathbf{x}_p}\right). \qquad [2.39]$$

hence we actually deduce the skew symmetry of the function $\psi_p(\mathbf{x}_1,\cdots,\mathbf{x}_p)$.

Justification of multi-linearity. Multi-linearity is now obtained immediately, since by definition we have:

$$\lambda[A,\mathbf{X}] \mathrm{R} \beta[A,\mathbf{X}] \Rightarrow J(\lambda[A,\mathbf{X}]) = J(\beta[A,\mathbf{X}]). \qquad [2.40]$$

Now, if we develop this expression by exploiting the linearity of operator J, we have:

$$J(\beta[A,\mathbf{X}]) = \frac{1}{p!} \sum_{i_1,\cdots,i_p} \alpha_{i_1}^1 \cdots \alpha_{i_p}^p \sum_{\sigma\in S_p} \varepsilon(\sigma) J\left(\lambda_{\sigma(\mathbf{x}_1,\cdots,\mathbf{x}_p)}\right), \qquad [2.41]$$

whereas according to the skew symmetry property, we obtain that:

$$\forall\sigma\in S_p,\ J\left(\lambda_{\sigma(\mathbf{x}_1,\cdots,\mathbf{x}_p)}\right) = \varepsilon(\sigma) J\left(\lambda_{\mathbf{x}_1,\cdots,\mathbf{x}_p}\right), \qquad [2.42]$$

hence we deduce:

$$\begin{aligned} J(\beta[A,\mathbf{X}]) &= \frac{1}{p!} \sum_{i_1,\cdots,i_p} \alpha_{i_1}^1 \cdots \alpha_{i_p}^p \sum_{\sigma\in S_p} J\left(\lambda_{\mathbf{x}_1,\cdots,\mathbf{x}_p}\right) \\ &= \sum_{i_1,\cdots,i_p} \alpha_{i_1}^1 \cdots \alpha_{i_p}^p J\left(\lambda_{\mathbf{x}_1,\cdots,\mathbf{x}_p}\right). \end{aligned} \qquad [2.43]$$

Considering the notations of function ψ_p, we have:

$$\psi_p\left(\sum_{i_1} \alpha_{i_1}^1 \mathbf{x}_{i_1}^1,\cdots,\sum_{i_p} \alpha_{i_p}^p \mathbf{x}_{i_p}^p\right) = \sum_{i_1,\cdots,i_p} \alpha_{i_1}^1 \cdots \alpha_{i_p}^p \psi_p\left(\mathbf{x}_{i_1}^1,\cdots,\mathbf{x}_{i_p}^p\right), \qquad [2.44]$$

which corresponds exactly to the multi-linearity of function ψ_p (see proposition 1.12).

2.2. Uniqueness of exterior algebra with degree p

THEOREM 2.2 (uniqueness of exterior algebra).– The pair $(\boldsymbol{F}, \psi_p)$ in the theorem of existence is "unique" up to isomorphism. In other words, if $(\boldsymbol{F}, \psi_p)$ and $(\boldsymbol{F}', \gamma_p)$ are two pairs that satisfy the definition of theorem 2.1, then there exists a unique invertible linear map $l : \boldsymbol{F} \mapsto \boldsymbol{F}'$ such that $\gamma_p = l \circ \psi_p$.

PROOF.– Let us start from the pair $(\boldsymbol{F}, \psi_p)$. Since $(\boldsymbol{F}', \gamma_p)$ is such that γ_p is a skew-symmetric map from $\boldsymbol{E}^p$ into $\boldsymbol{F}'$, there exists a unique linear map $\alpha : \boldsymbol{F} \mapsto \boldsymbol{F}'$ such that we have: $\gamma_p = \alpha \circ \psi_p$. Starting this time from the pair $(\boldsymbol{F}', \gamma_p)$ there exists a unique linear map $\beta : \boldsymbol{F}' \mapsto \boldsymbol{F}$ such that $\psi_p = \beta \circ \gamma_p$. By composition, there exists a unique linear map $\beta \circ \alpha$ such that $\psi_p = \beta \circ \alpha \circ \psi_p$ and a unique linear map $\alpha \circ \beta$ such that $\gamma_p = \alpha \circ \beta \circ \gamma_p$. Moreover, since we clearly have $\psi_p = Id_{\boldsymbol{F}} \circ \psi_p$ and $\gamma_p = Id_{\boldsymbol{F}'} \circ \gamma_p$, we therefore deduce, in virtue of uniqueness:

$$\alpha \circ \beta = Id_{\boldsymbol{F}'}, \quad \beta \circ \alpha = Id_{\boldsymbol{F}}, \qquad [2.45]$$

which shows that the map α is both injective and surjective from $\boldsymbol{F}$ into $\boldsymbol{F}'$ and thus it is an isomorphism (since it is linear). Its inverse is then the map β. Therefore, the theorem is demonstrated with $l = \alpha$.

We can now introduce some vocabulary regarding exterior algebras.

DEFINITION 2.1.– Let $\boldsymbol{E}$ be a linear space:

1) For all $p \geq 2$ and for any pair $(\boldsymbol{F}, \psi_p)$ of theorem 2.1, we say that $\boldsymbol{F}$ is a **p-exterior algebra** over $\boldsymbol{E}$ and that ψ_p is **a construction operator** of the algebra. The integer p is called the degree of the exterior algebra.

2) For $p \geq 2$, since p-exterior algebras are all isomorphic (according to the previous theorem), we denote by $\Lambda^p \boldsymbol{E}$ the class (of equivalence) of p-exterior algebras over $\boldsymbol{E}$. In practice, we shall make no distinction between the equivalence class and any of its representatives.

3) The elements of the exterior algebra $\Lambda^p \boldsymbol{E}$ are called p-vectors.

4) The elements of $\Lambda^p \boldsymbol{E}$ written in the form $\psi_p(\mathbf{x}_1, \cdots, \mathbf{x}_p)$ are called p-blades.

2.3. Some remarks

Let us linger over the following points:

1) The definition of exterior algebra by means of characterization 2.1 still works when $p = 1$. In fact, it suffices to set $\boldsymbol{F} = \boldsymbol{E}$, and to choose ψ_1 the identity map from

$\boldsymbol{E}$ into $\boldsymbol{E}$ (which is trivially skew-symmetric from $\boldsymbol{E}$ to $\boldsymbol{E}$). We then naturally obtain $\Lambda^1\boldsymbol{E} = \boldsymbol{E}$.

2) However, it is not possible to generalize the notion of exterior algebra of degree p when $p = 0$. In all that follows, we shall thus set in an axiomatic way:

DEFINITION 2.2.– For any linear space $\boldsymbol{E}$, we use the notation $\Lambda^0\boldsymbol{E} := \mathbb{R}$.

In this case, a constructor $\psi_0 : \mathbb{R} \mapsto \mathbb{R}$ is the identity operator.

3) Since the map ψ_p with values in $\boldsymbol{F}$ is p-linear (and not linear!), the classical results regarding images of linear operators cannot be applied. In particular, the image $\psi_p(\boldsymbol{E}^p)$ is not necessarily a linear space.

4) The consequence is that in general we have $\psi_p(\boldsymbol{E}^p) \subset \boldsymbol{F}$, $\psi_p(\boldsymbol{E}^p) \neq \boldsymbol{F}$ and thus potentially there exist elements of $\boldsymbol{F}$ that cannot be written as an image of map ψ_p.

5) It is not so immediate to construct a basis of $\boldsymbol{F}$ starting from a basis of $\boldsymbol{E}$ and a map ψ_p, as it is often the case for linear spaces constructed from other linear spaces (as well as from the Cartesian product of linear spaces). In fact, this issue can be easily understood by considering the fact that the process of constructing exterior spaces has employed non-constructive theorems regarding bases, using the Zorn lemma. The identification of exterior algebra bases will require a fundamental detour that consists of constructing the exterior product's symbol.

6) We could have constructed the exterior algebra with degree p from a tensor algebra $\mathcal{T}^p\boldsymbol{E}$ constructed over $\boldsymbol{E}$. The construction of the tensor algebra follows exactly the same method that we employed, although it changes the equivalence relation defined over $\mathcal{F}_0(\boldsymbol{E}^p \mapsto \mathbb{R})$ by defining instead the following set $\mathcal{G}$:

$$\mathcal{G} = \lambda_{\sum_{i_1}\alpha^1_{i_1}\mathbf{x}^1_{i_1},\cdots,\sum_{i_p}\alpha^p_{i_p}\mathbf{x}^p_{i_p}} - \sum_{i_1,\cdots,i_k}\alpha^1_{i_1}\cdots\alpha^p_{i_p}\lambda_{\mathbf{x}^1_{i_1},\cdots,\mathbf{x}^p_{i_p}} \qquad [2.46]$$

we then have a construction operator ϕ_p that maps $\boldsymbol{E}^p$ into the quotient space $\mathcal{F}_0(\boldsymbol{E}^p \mapsto \mathbb{R}) / \langle\mathcal{G}\rangle$ and which is multi-linear (but not skew-symmetric). Then, once the tensor algebra is constructed, we can define a subset G within this space according to:

$$G = \left\{\phi_p(\mathbf{x}_1,\cdots,\mathbf{x}_p) - \sum_{\sigma\in S_p}\varepsilon(\sigma)\,\phi_p\left(\mathbf{x}_{\sigma(1)},\cdots,\mathbf{x}_{\sigma(p)}\right)\right\} \qquad [2.47]$$

when $\mathbf{x}_1,\cdots,\mathbf{x}_p$ spans $\boldsymbol{E}^p$ and we consider the quotient set $\mathcal{T}^p\boldsymbol{E}/\langle G\rangle$ where $\langle G\rangle$ is the linear space generated by G. This set is an exterior algebra with degree p, and the composition of the canonical surjection from $\mathcal{T}^p\boldsymbol{E}$ into $\mathcal{T}^p\boldsymbol{E}/\langle G\rangle$ and the operator ϕ_p from $\boldsymbol{E}^p$ to $\mathcal{T}^p\boldsymbol{E}$ is a constructor ψ_p of the exterior algebra.

3

Exterior Product Symbol

One of the most important questions regarding vector families concerns the independence or the relation among them:

QUESTION 3.1.– Let $\mathbf{x}_1, \cdots, \mathbf{x}_p$ be a family of elements of a linear space $\boldsymbol{E}$. Is it possible to characterize whether this family is independent or not?

We shall see of course that it is indeed possible, provided that a (formal) calculation tool is created, which consists of the exterior product of vectors.

3.1. Exterior product symbol $\wedge$ in exterior algebras

3.1.1. *Construction principle*

THEOREM 3.1.– Let $\boldsymbol{E}$ be a linear space and p, q be two integers. We denote, respectively, by $(\Lambda^p \boldsymbol{E}, \psi_p)$, $(\Lambda^q \boldsymbol{E}, \psi_q)$ and by $(\Lambda^{p+q} \boldsymbol{E}, \psi_{p+q})$, the exterior algebras with successive degrees $p, q, p+q$ equipped with their associated constructors. Then, there exists a unique bilinear map $B : \Lambda^p \boldsymbol{E} \times \Lambda^q \boldsymbol{E} \mapsto \Lambda^{p+q} \boldsymbol{E}$ such that the following equality is satisfied:

$$\forall \mathbf{x}_1, \cdots, \mathbf{x}_p, \forall \mathbf{y}_1, \cdots, \mathbf{y}_q, \; B\left(\psi_p\left(\mathbf{x}_1, \cdots, \mathbf{x}_p\right), \psi_q\left(\mathbf{y}_1, \cdots, \mathbf{x}_q\right)\right)$$
$$= \psi_{p+q}\left(\mathbf{x}_1, \cdots, \mathbf{x}_p, \mathbf{y}_1, \cdots, \mathbf{y}_q\right). \qquad [3.1]$$

PROOF.– Let us suppose in fact that the vectors $\mathbf{x}_1, \cdots, \mathbf{x}_p$ are fixed, then the map $M[\mathbf{x}_1, \cdots, \mathbf{x}_p](\cdot)$ from $\boldsymbol{E}^q$ with values in $\Lambda^{p+q}\boldsymbol{E}$ defined as:

$$\forall \mathbf{y}_1, \cdots \mathbf{y}_q \in \boldsymbol{E}^q, M[\mathbf{x}_1, \cdots, \mathbf{x}_p](\mathbf{y}_1, \cdots \mathbf{y}_q) := \psi_{p+q}(\mathbf{x}_1, \cdots, \mathbf{x}_p, \mathbf{y}_1, \cdots, \mathbf{y}_q) \quad [3.2]$$

is a map which is clearly skew-symmetric and q-linear over $\boldsymbol{E}^q$. As a result, according to the construction theorem 2.1, there exists a unique linear map $l[\mathbf{x}_1, \cdots, \mathbf{x}_p](\cdot)$ from $\Lambda^q\boldsymbol{E} \mapsto \Lambda^{p+q}\boldsymbol{E}$ such that we have the following equality:

$$l[\mathbf{x}_1, \cdots, \mathbf{x}_p](\psi_q(\mathbf{y}_1, \cdots \mathbf{y}_q)) = \psi_{p+q}(\mathbf{x}_1, \cdots, \mathbf{x}_p, \mathbf{y}_1, \cdots, \mathbf{y}_q). \quad [3.3]$$

Thanks to the result above, it is possible to construct, for any family $\mathbf{x}_1, \cdots, \mathbf{x}_p$, a linear map over $\Lambda^q\boldsymbol{E}$ with values in $\Lambda^{p+q}\boldsymbol{E}$, i.e. an element of $\mathcal{L}(\Lambda^q\boldsymbol{E}, \Lambda^{p+q}\boldsymbol{E})$. It is very clear that for all permutation $\sigma \in S_p$ and for any family $\mathbf{y}_1, \cdots, \mathbf{y}_q$, we have:

$$l(\mathbf{x}_{\sigma(1)}, \cdots, \mathbf{x}_{\sigma(p)})(\psi_q(\mathbf{y}_1, \cdots \mathbf{y}_q)) = \varepsilon(\sigma)\psi_{p+q}(\mathbf{x}_1, \cdots, \mathbf{x}_p, \mathbf{y}_1, \cdots, \mathbf{y}_q). \quad [3.4]$$

In particular, this shows that we have:

$$l(\mathbf{x}_{\sigma(1)}, \cdots, \mathbf{x}_{\sigma(p)})(\psi_q(\mathbf{y}_1, \cdots \mathbf{y}_q)) = \varepsilon(\sigma) l[\mathbf{x}_1, \cdots, \mathbf{x}_p](\psi_q(\mathbf{y}_1, \cdots \mathbf{y}_q)). \quad [3.5]$$

Therefore, we deduce, by means of uniqueness of linear operators, that we have:

$$l[\mathbf{x}_{\sigma(1)}, \cdots, (\mathbf{x}_{\sigma(p)}](\cdot) = \varepsilon(\sigma) l[\mathbf{x}_1, \cdots, \mathbf{x}_p](\cdot), \quad [3.6]$$

with the equality within the linear map. In other words, the map:

$$\mathbf{x}_1, \cdots, \mathbf{x}_p \in \boldsymbol{E}^p \mapsto l[\mathbf{x}_1, \cdots, \mathbf{x}_p](\cdots) \in \mathcal{L}(\Lambda^q\boldsymbol{E}, \Lambda^{p+q}\boldsymbol{E}) \quad [3.7]$$

is a map which is (clearly) p-linear but also skew-symmetric. As a result, according to the construction theorem 2.1, there exists a unique linear map $L : \Lambda^p\boldsymbol{E} \mapsto \mathcal{L}(\Lambda^q\boldsymbol{E}, \Lambda^{p+q}\boldsymbol{E})$ such that:

$$L(\psi_p(\mathbf{x}_1, \cdots, \mathbf{x}_p))(\cdot) = l[\mathbf{x}_1, \cdots, \mathbf{x}_p](\cdots), \quad [3.8]$$

since in general the linear space of bilinear forms $\mathcal{M}_2(\boldsymbol{X} \times \boldsymbol{Y}, \boldsymbol{Z})$ is isomorphic to the linear space $\mathcal{L}(\boldsymbol{X}, \mathcal{L}(\boldsymbol{Y}, \boldsymbol{Z}))$ (see proposition 1.13), we deduce thereof what was required: there exists a unique bilinear form $B : \Lambda^p\boldsymbol{E} \times \Lambda^q\boldsymbol{E} \mapsto \Lambda^{p+q}\boldsymbol{E}$ such that:

$$B(\psi_p(\mathbf{x}_1, \cdots, \mathbf{x}_p), \psi_q(\mathbf{y}_1, \cdots, \mathbf{y}_q)) = \psi_{p+q}(\mathbf{x}_1, \cdots, \mathbf{x}_p, \mathbf{y}_1, \cdots, \mathbf{y}_q). \quad [3.9]$$

REMARK 3.1.– *A priori*, the construction of the operator B is not proven if one among the integers p or q is zero. ■

DEFINITION 3.1.– Given the constructors $\psi_p, \psi_q, \psi_{p+q}$ of the spaces $\Lambda^p \boldsymbol{E}$, $\Lambda^q \boldsymbol{E}$ and $\Lambda^{p+q} \boldsymbol{E}$ we define the exterior product symbol according to the notation:

$$\forall \mathbf{E}_p \in \Lambda^p \boldsymbol{E},\ \forall \mathbf{E}_q \in \Lambda^q \boldsymbol{E},\ \ \mathbf{E}_p \wedge \mathbf{E}_q := B\left(\mathbf{E}_p, \mathbf{E}_q\right), \qquad [3.10]$$

where B is the bilinear form of theorem 3.1. If either p or q is zero, we define then:

$$\forall \lambda \in \mathbb{R},\ \forall \mathbf{X} \in \Lambda^p \boldsymbol{E},\ \ \lambda \wedge \mathbf{X} := \mathbf{X} \wedge \lambda := \lambda \mathbf{X}, \qquad [3.11]$$

where, in the last equality, the product represents the action of a scalar on a vector.

A priori, the exterior product symbol depends on the constructors that were chosen for the exterior algebras. In practice, since all the constructors are isomorphic to each other (according to the uniqueness theorem of an exterior algebra 2.2), we deduce thereof that all the exterior product symbols are isomorphic to each other according to the following proposition:

PROPOSITION 3.1.– Let $\psi_p, \psi_q, \psi_{p+q}$ and $\psi'_p, \psi'_q, \psi'_{p+q}$ be two families of constructors of spaces $\Lambda^p \boldsymbol{E}$, $\Lambda^q \boldsymbol{E}$ and $\Lambda^{p+q} \boldsymbol{E}$ that entail the exterior product symbols $\wedge, \wedge'$. Then, there exist some isomorphisms L_p, L_q and L_{p+q} over $\Lambda^p \boldsymbol{E}$, $\Lambda^q \boldsymbol{E}$ and $\Lambda^{p+q} \boldsymbol{E}$ such that we have:

$$\forall \mathbf{E}_p \in \Lambda^p \boldsymbol{E},\ \forall \mathbf{E}_q \in \Lambda^q \boldsymbol{E},\ \ \mathbf{E}_p \wedge \mathbf{E}_q = L_{p+q}^{-1}\left(L_p\left(\mathbf{E}_p\right) \wedge' L_q\left(\mathbf{E}_q\right)\right). \qquad [3.12]$$

PROOF.– By construction, we know that:

$$B'\left(\psi'_p, \psi'_q\right) = \psi'_{p+q},\ \ B\left(\psi_p, \psi_q\right) = \psi_{p+q}. \qquad [3.13]$$

Now, given the uniqueness of exterior algebra, we know that there exists a unique isomorphism L_{p+q} over $\Lambda^{p+q} \boldsymbol{E}$ such that we have $\psi'_{p+q} = L_p \circ \psi_p$. The same is true for the functions ψ'_p, ψ_p, and ψ'_q, ψ_q over $\Lambda^p \boldsymbol{E}$ and $\Lambda^q \boldsymbol{E}$. We can thus write that:

$$B\left(\psi_p, \psi_q\right) = L_{p+q}^{-1} \circ B'\left(L_p \circ \psi_p, L_q \circ \psi_q\right) = \psi_{p+q}. \qquad [3.14]$$

In virtue of uniqueness of bilinear form in theorem 3.1, we hence deduce the equality:

$$B\left(\cdot, \cdot\right) = L_{p+q}^{-1} \circ B'\left(L_p\left(\cdot\right), L_q\left(\cdot\right)\right) \qquad [3.15]$$

over the product set $\Lambda^p \boldsymbol{E} \times \Lambda^q \boldsymbol{E}$. This is exactly what we were seeking to demonstrate.

3.1.2. *Properties*

PROPOSITION 3.2.– The exterior product symbol satisfies the following properties:

1) The symbol "$\wedge$" is an associative symbol.

2) The symbol "$\wedge$" is bilinear: thus, it defines a product operation over $\Lambda^p \boldsymbol{E} \times \Lambda^q \boldsymbol{E}$ with values in $\Lambda^{p+q} \boldsymbol{E}$ that is both right- and left-distributive over addition.

3) For all $\mathbf{E}_p \in \Lambda^p \boldsymbol{E}$ and all $\mathbf{E}_q \in \Lambda^q \boldsymbol{E}$, we have the formula:

$$\mathbf{E}_1 \wedge \mathbf{E}_2 = (-1)^{pq}\, \mathbf{E}_q \wedge \mathbf{E}_p. \qquad [3.16]$$

4) Applied to any family $\mathbf{x}_1, \cdots, \mathbf{x}_p$ of a space $\boldsymbol{E}$, we have by definition:

$$\mathbf{x}_1 \wedge \cdots \wedge \mathbf{x}_p = \psi_p\left(\mathbf{x}_1, \cdots, \mathbf{x}_p\right), \qquad [3.17]$$

therefore the symbol $\wedge$ defines an alternating p-linear map of $\boldsymbol{E}^p$ with values in $\Lambda^p \boldsymbol{E}$.

PROOF.– Let us demonstrate these properties:

1) Let us choose three families $\mathbf{x}_1, \cdots, \mathbf{x}_p$, as well as $\mathbf{y}_1, \cdots, \mathbf{y}_q$ and $\mathbf{z}_1, \cdots, \mathbf{z}_r$, of vectors in $\boldsymbol{E}$. Then, it is clear that the following expression:

$$\left[\psi_p\left(\mathbf{x}_1, \ldots, \mathbf{x}_p\right) \wedge \psi_q\left(\mathbf{y}_1, \ldots, \mathbf{y}_p\right)\right] \wedge \psi_r\left(\mathbf{z}_1, \ldots, \mathbf{z}_r\right) \qquad [3.18]$$

is equal (by construction of the symbol $\wedge$ of the term within the parentheses) to the expression:

$$\psi_{p+q}\left(\mathbf{x}_1, \ldots, \mathbf{x}_p, \mathbf{y}_1, \ldots, \mathbf{y}_p\right) \wedge \psi_r\left(\mathbf{z}_1, \ldots, \mathbf{z}_r\right), \qquad [3.19]$$

which, in its turn, is equal (by construction of the symbol $\wedge$) to:

$$\psi_{p+q+r}\left(\mathbf{x}_1, \ldots, \mathbf{x}_p, \mathbf{y}_1, \ldots, \mathbf{y}_p, \mathbf{z}_1, \ldots, \mathbf{z}_r\right). \qquad [3.20]$$

In the same way, it can be shown that the term:

$$\psi_p\left(\mathbf{x}_1, \ldots, \mathbf{x}_p\right) \wedge \left[\psi_q\left(\mathbf{y}_1, \ldots, \mathbf{y}_p\right) \wedge \psi_r\left(\mathbf{z}_1, \ldots, \mathbf{z}_r\right)\right] \qquad [3.21]$$

is also equal to the previous term, namely to:

$$\psi_{p+q+r}\left(\mathbf{x}_1, \ldots, \mathbf{x}_p, \mathbf{y}_1, \ldots, \mathbf{y}_p, \mathbf{z}_1, \ldots, \mathbf{z}_r\right), \qquad [3.22]$$

hence we easily deduce the associativity of the exterior product.

2) By definition, the symbol $\wedge$ is defined by means of a bilinear map (see theorem 3.1). Hence, the property of bilinearity.

3) It suffices to demonstrate that we have:

$$\psi_p(\mathbf{x}_1,\cdots,\mathbf{x}_p)\wedge\psi_q(\mathbf{y}_1,\cdots,\mathbf{y}_q) = (-1)^{pq}\,\psi_q(\mathbf{y}_1,\cdots,\mathbf{y}_q)\wedge\psi_p(\mathbf{x}_1,\cdots,\mathbf{x}_p). \quad [3.23]$$

Thus, the bilinear map $B_{pq}: \Lambda^p\boldsymbol{E}\times\Lambda^q\boldsymbol{E}\mapsto\Lambda^{p+q}\boldsymbol{E}$ will be related to the map $B_{qp}:\Lambda^p\boldsymbol{E}\times\Lambda^q\boldsymbol{E}\mapsto\Lambda^{p+q}\boldsymbol{E}$ according to the following expression:

$$B_{pq}(\psi_p,\psi_q) = (-1)^{pq}\,B_{qp}(\psi_q,\psi_p). \quad [3.24]$$

In virtue of the construction uniqueness theorem of bilinear forms that are coincident for the elements ψ_p,ψ_q (see 3.1), this will show that we have indeed $B_{pq}: \Lambda^p\boldsymbol{E}\times\Lambda^q\boldsymbol{E}\mapsto\Lambda^{p+q}\boldsymbol{E}$ equal to $(-1)^{pq}\,B_{qp}:\Lambda^q\boldsymbol{E}\times\Lambda^p\boldsymbol{E}\mapsto\Lambda^{p+q}\boldsymbol{E}$ and thus that we have indeed, for any pair $\mathbf{E}_p\in\Lambda^p\boldsymbol{E}$ and $\mathbf{E}_q\in\Lambda^q\boldsymbol{E}$, the relation:

$$\mathbf{E}_p\wedge\mathbf{E}_q = (-1)^{pq}\,\mathbf{E}_q\wedge\mathbf{E}_p. \quad [3.25]$$

By definition, we know that:

$$\psi_p(\mathbf{x}_1,\cdots,\mathbf{x}_p)\wedge\psi_q(\mathbf{y}_1,\cdots,\mathbf{y}_q) = \psi_{p+q}(\mathbf{x}_1,\cdots,\mathbf{x}_p,\mathbf{y}_1,\cdots,\mathbf{y}_q) \quad [3.26]$$

now, in order to move from $\psi_{p+q}(\mathbf{x}_1,\cdots,\mathbf{x}_p,\mathbf{y}_1,\cdots,\mathbf{y}_q)$ to the term $\psi_{p+q}(\mathbf{y}_1,\cdots,\mathbf{y}_q,\mathbf{x}_1,\cdots,\mathbf{x}_p)$ which satisfies precisely:

$$\psi_{p+q}(\mathbf{y}_1,\cdots,\mathbf{y}_q,\mathbf{x}_1,\cdots,\mathbf{x}_p) = \psi_q(\mathbf{y}_1,\cdots,\mathbf{y}_q)\wedge\psi_p(\mathbf{x}_1,\cdots,\mathbf{x}_p) \quad [3.27]$$

the following permutation $\sigma\in S_{p+q}$ has to be performed: $\sigma(1)=q+1,\cdots,\sigma(p)=p+q$ then $\sigma(p+1)=1$ up to $\sigma(p+q)=q$. This process requires a number of exactly pq transpositions: we consider element $p+1$ and we exchange it, step by step, until it reaches element 1. This requires p transpositions. Then, we consider the element $p+2$ and we exchange it, step by step, until it reaches element 2: again, this requires p transpositions. And so forth, until element $p+q$ is moved to the element p. Overall, pq transpositions are performed. Hence the result, since the signature of a transposition is equal to -1.

4) Evident, since the constructor is essentially a skew-symmetric p-linear map (see construction theorem 2.1).

3.1.3. *Grassmann's algebra*

DEFINITION 3.2.– Let $\boldsymbol{E}$ be a linear space. Consider the Cartesian product of a (infinite countable) set of exterior algebras:

$$\Lambda^0 \boldsymbol{E} \times \Lambda^1 \boldsymbol{E} \times \Lambda^2 \boldsymbol{E} \times \cdots. \qquad [3.28]$$

A *multi-vector* over $\boldsymbol{E}$ is then any element of the form:

$$G := (G_0, \mathbf{G}_1, \mathbf{G}_2, \cdots), \qquad [3.29]$$

where only a finite number of components $\mathbf{G}_i, i \in \mathbb{N}$ are non-zero. We denote by $\mathcal{G}[\boldsymbol{E}]$ the set of multi-vectors over $\boldsymbol{E}$.

It is clear that we can equip $\mathcal{G}[\boldsymbol{E}]$ with a linear space structure according to:

$$G + H := (G_0 + H_0, \mathbf{G}_1 + \mathbf{H}_1, \mathbf{G}_2 + \mathbf{H}_2, \cdots) \qquad [3.30]$$

$$\lambda \mathbf{G} := (\lambda G_0, \lambda \mathbf{G}_1, \lambda \mathbf{G}_2, \cdots) \qquad [3.31]$$

However, since we have introduced a product between elements of exterior algebras, we can employ it to define a product between multi-vectors by introducing a convolution:

DEFINITION 3.3.– Let $G := (G_0, \mathbf{G}_1, \mathbf{G}_2, \cdots)$ and $H := (H_0, \mathbf{H}_1, \mathbf{H}_2, \cdots)$ be two elements of $\mathcal{G}[\boldsymbol{E}]$. We define their Grassmann's product, again denoted by "$\wedge$", by the convolution formula:

$$G \wedge H = \sum_{p=0}^{p=+\infty} \left[\sum_{k=0}^{k=p} \mathbf{G}_k \wedge \mathbf{H}_{p-k} \right]. \qquad [3.32]$$

The definition is consistent due to the fact that in the summation, which is infinite *a priori*, only a finite number of terms are non-zero. The following properties can be easily verified:

PROPOSITION 3.3.– Equipped with its addition and exterior product, Grassmann's algebra $(\mathcal{G}(\boldsymbol{E}), +, \wedge)$ has indeed a structure of a unitary non-commutative algebra:

1) the set $(\mathcal{G}(\boldsymbol{E}), +)$ has the structure of a linear space;

2) the operation $\wedge$ is an associative product;

3) the product $\wedge$ is right- and left-distributive over addition;

4) the product admits a neutral element, the real 1.

Note that right- and left-distributivity of the exterior product are simply due to the fact that the latter defines a multi-linear operation over its arguments.

DEFINITION 3.4.– Let $G \in \mathcal{G}$ be a multi-vector written in the form $G = (G_0, \mathbf{G}_1, \mathbf{G}_2, \cdots)$:

1) We say that it is homogeneous of degree p if $\forall i \neq p, \mathbf{G}_i = \mathbf{0}$;

2) The map $\{\cdot\}_p : \mathcal{G}(\boldsymbol{E}) \mapsto \Lambda^p \boldsymbol{E}$ defined as:

$$\forall G = (G_0, \mathbf{G}_1, \mathbf{G}_2, \cdots), \ \{G\}_p := \mathbf{G}_p \qquad [3.33]$$

is a p-grade operator.

From the structural point of view, it can be summarized as:

PROPOSITION 3.4.– The set $\mathcal{G}[E]$ equipped with its linear space structure and with (Grassmann's) product admits the structure of a graded algebra.

Let us recall that the structure of an algebra is graded when:

1) the algebra admits a direct sum decomposition of the form $\mathcal{G} = \otimes_{p \in \mathbb{N}} \mathcal{G}_p$ where each $\mathcal{G}_p$ is a linear subspace of $\mathcal{G}$;

2) the product maps $\mathcal{G}_p \times \mathcal{G}_q$ into $\mathcal{G}_{p+q}$ (which is simply the result of the definition of exterior product between p and q vectors).

3.2. Symbol of exterior product $\wedge^*$ between forms

3.2.1. *Construction principle*

PROPOSITION 3.5.– Let $l_p : \Lambda^p \boldsymbol{E} \mapsto \mathbb{R}$ and $l_q : \Lambda^q \boldsymbol{E} \mapsto \mathbb{R}$ be two linear maps with values in $\mathbb{R}$ (linear forms over $\Lambda^p \boldsymbol{E}$ and over $\Lambda^q \boldsymbol{E}$). Then, the map defined over $\boldsymbol{E}^{p+q}$ by the relation:

$$\gamma(\mathbf{x}_1, \cdots, \mathbf{x}_{p+q}) := \sum_{\sigma \in S_{p+q}} \frac{\varepsilon(\sigma)}{(p+q)!} l_p\left[\psi_p\left(\mathbf{x}_{\sigma(1)}, \cdots, \mathbf{x}_{\sigma(p)}\right)\right]$$

$$l_q\left[\psi_q\left(\mathbf{x}_{\sigma(p+1)}, \cdots, \mathbf{x}_{\sigma(p+q)}\right)\right] \quad [3.34]$$

is a $(p+q)$ skew-symmetric form.

PROOF.– The fact that the map is $p+q$-linear simply stems from the maps $l_p \circ \psi_p$ and $l_q \circ \psi_q$ being respectively p, q linear. Let $\alpha \in S_{p+q}$ be a permutation of $[1, p+q]$. We calculate $\gamma\left(\mathbf{x}_{\alpha(1)}, \cdots, \mathbf{x}_{\alpha(p+q)}\right)$ according to:

$$\sum_{\sigma \in S_{p+q}} \frac{\varepsilon(\sigma)}{(p+q)!} l_p\left[\psi_p\left(\mathbf{x}_{\sigma(\alpha(1))}, \cdots, \mathbf{x}_{\sigma(\alpha(p))}\right)\right]$$

$$l_q\left[\psi_q\left(\mathbf{x}_{\sigma(\alpha(p+1))}, \cdots, \mathbf{x}_{\sigma(\alpha(p+q))}\right)\right], \quad [3.35]$$

we write then $\sigma = \sigma \circ \alpha \circ \alpha^{-1}$ and we have thus (see theorem 1.4):

$$\varepsilon(\sigma) = \varepsilon\left(\sigma \circ \alpha \circ \alpha^{-1}\right) = \varepsilon\left(\alpha^{-1}\right)\varepsilon(\sigma \circ \alpha) = \varepsilon(\alpha)\varepsilon(\sigma \circ \alpha). \quad [3.36]$$

In particular, the term $\gamma\left(\mathbf{x}_{\alpha(1)}, \cdots, \mathbf{x}_{\alpha(p+q)}\right)$ is equal to:

$$\varepsilon(\alpha) \sum_{\sigma \in S_{p+q}} \frac{\varepsilon(\sigma \circ \alpha)}{(p+q)!} l_p\left[\psi_p\left(\mathbf{x}_{\sigma(\alpha(1))}, \cdots, \mathbf{x}_{\sigma(\alpha(p))}\right)\right]$$

$$l_q\left[\psi_q\left(\mathbf{x}_{\sigma(\alpha(p+1))}, \cdots, \mathbf{x}_{\sigma(\alpha(p+q))}\right)\right]. \quad [3.37]$$

By performing the "change of variable" $\sigma \leftrightarrow \sigma \circ \alpha$ inside the summation over the permutations, we obtain:

$$\gamma\left(\mathbf{x}_{\alpha(1)}, \cdots, \mathbf{x}_{\alpha(p+q)}\right) = \varepsilon(\alpha)\gamma(\mathbf{x}_1, \cdots, \mathbf{x}_{p+q}), \quad [3.38]$$

which proves that the p-linear is skew-symmetric.

DEFINITION 3.5.– According to the theorem of exterior algebra construction (see theorem 2.1), there exists a unique linear form $l : \Lambda^{p+q}\boldsymbol{E} \mapsto \mathbb{R}$ such that we have $l \circ \psi_{p+q} = \gamma$. We set by definition:

$$l := l_1 \wedge^* l_2. \qquad [3.39]$$

3.2.2. *Elementary properties of the symbol of exterior product between forms*

PROPOSITION 3.6.– The following properties are true:

1) The symbol of exterior product $\wedge^*$ is associative:

$$\forall l_p \in (\Lambda^p \boldsymbol{E})^*, \forall l_q (\Lambda^q \boldsymbol{E})^*, \forall l_r (\Lambda^r \boldsymbol{E})^*, (l_p \wedge^* l_q) \wedge^* l_r = l_p \wedge^* (l_q \wedge^* l_r). \qquad [3.40]$$

2) We can then define the exterior product of a family of linear forms. The symbol "$\wedge^*$" defines then a multi-linear map:

$$(\Lambda^{i_1}\boldsymbol{E})^* \times \cdots \times (\Lambda^{i_p}\boldsymbol{E})^* \mapsto (\Lambda^{i_1+\cdots+i_p}\boldsymbol{E})^*. \qquad [3.41]$$

3) If $l_p \in (\Lambda^p \boldsymbol{E})^*$ and if $l_q \in (\Lambda^q \boldsymbol{E})^*$, then we have $l_p \wedge^* l_q = (-1)^{pq} l_q \wedge^* l_p$.

4) The symbol "$\wedge^*$" enables the definition of a map:

$$\boldsymbol{E}^* \times \cdots \times \boldsymbol{E}^* \mapsto (\Lambda^p \boldsymbol{E})^* : (\phi_1, \cdots, \phi_p) \mapsto \phi_1 \wedge^* \cdots \wedge^* \phi_p, \qquad [3.42]$$

which is skew-symmetric p-linear.

PROOF.–

1) Let us calculate:

$$\sum_{\sigma \in S_{p+q+r}} \frac{\varepsilon(\sigma)}{(p+q+r)!} l_p(\mathbf{x}_{\sigma(1)}, \cdots) l_q(\mathbf{x}_{\sigma(p+1)}, \cdots) l_r(\mathbf{x}_{\sigma(p+q+1)}, \cdots). \qquad [3.43]$$

By setting $\alpha \in S_{p+q}$, let us consider the permutation $\beta \in S_{p+q+r}$ defined by $\beta = (\alpha, Id)$. It is clear that we have $\sigma \in S_{p+q+r}$ if and only if $\beta \circ \sigma \in S_{p+q+r}$. By means of a "change of variables" $\sigma \leftrightarrow \beta \circ \sigma$, we hence deduce that the above term can be calculated as:

$$\sum_{\sigma \in S_{p+q+r}} \frac{\varepsilon(\sigma)\varepsilon(\alpha)}{(p+q+r)!} l_p\left(\mathbf{x}_{\alpha\circ\sigma(1)}, \cdots\right)$$
$$l_q\left(\mathbf{x}_{\alpha\circ\sigma(p+1)}, \cdots\right) l_r\left(\mathbf{x}_{\sigma(p+q+1)}, \cdots\right), \quad [3.44]$$

since the result is true for any permutation $\alpha \in S_{p+q}$, independently of the permutation, it is possible to perform a summation over all $\alpha \in S_{p+q}$ dividing by $(p+q)!$. We obtain then:

$$\sum_{\sigma \in S_{p+q+r}} \sum_{\alpha \in S_{p+q}} \frac{\varepsilon(\sigma)}{(p+q+r)!} \frac{\varepsilon(\alpha)}{(p+q)!} l_p\left(\mathbf{x}_{\alpha\circ\sigma(1)}, \cdots\right)$$
$$l_q\left(\mathbf{x}_{\alpha\circ\sigma(p+1)}, \cdots\right) l_r\left(\mathbf{x}_{\sigma(p+q+1)}, \cdots\right), \quad [3.45]$$

which can be rewritten as:

$$\sum_{\sigma \in S_{p+q+r}} \frac{1}{(p+q+r)!} \left(l_p \wedge^* l_q\right)\left(\mathbf{x}_{\sigma(1)}, \cdots, \mathbf{x}_{\sigma(p+q)}\right) l_r\left(\mathbf{x}_{\sigma(p+q+1)}, \cdots\right), \quad [3.46]$$

which by definition is equal to:

$$\left[\left(l_p \wedge^* l_q\right) \wedge^* l_r\right]\left(\mathbf{x}_1, \cdots, \mathbf{x}_{p+q+r}\right). \quad [3.47]$$

On the other hand, following a similar argumentation and considering the permutations of the form $\beta = (Id, \alpha)$, with α permutating over $[p+1, \cdots, p+q+r]$, it can be shown that the term:

$$\sum_{\sigma \in S_{p+q+r}} \frac{\varepsilon(\sigma)}{(p+q+r)!} l_p\left(\mathbf{x}_{\sigma(1)}, \cdots\right) l_q\left(\mathbf{x}_{\sigma(p+1)}, \cdots\right) l_r\left(\mathbf{x}_{\sigma(p+q+1)}, \cdots\right) \quad [3.48]$$

is calculated as:

$$\left[l_p \wedge^* \left(l_q \wedge^* l_r\right)\right]\left(\mathbf{x}_1, \cdots, \mathbf{x}_{p+q+r}\right), \quad [3.49]$$

hence the associativity result is deduced.

2) Associativity enables thus the definition of the exterior product symbol between linear forms for an arbitrary number of forms. Consider a natural integer k, as well as a family $i_1, \cdots, i_k$ of integers. We denote by $q = i_1 + \cdots + i_k$ and we consider k linear

forms $l_1, \cdots, l_k$ defined over $\Lambda^{i_m}\mathbf{E}$, $m \in [1, k]$ with values in $\mathbb{R}$. By generalization of the associativity calculation, we have:

$$[l_1 \wedge^* \cdots \wedge^* l_k] \circ \psi_q (\mathbf{x}_1, \cdots, \mathbf{x}_q)$$
$$= \sum_{\sigma \in S_q} \frac{\varepsilon(\sigma)}{q!} \pi_{m=1}^{m=k} \left[l_m \circ \psi_m \left(\mathbf{x}_{\sigma(i_{m-1}+1)}, \cdots, \mathbf{x}_{\sigma(i_m)}\right)\right] \quad [3.50]$$

with the convention $i_0 = 0$. In particular, by replacing any linear form $l_m \in \left(\Lambda^{i_m}\mathbf{E}\right)^*$ with a combination of the form $\alpha_m + \lambda\beta_m$ and by employing the product distributivity (of reals) over addition, we soon obtain a demonstration of the multi-linearity of symbol "$\wedge^*$".

3) We start from the formula that expresses $l_p \wedge l_q$ over elements of the form $\psi_{p+q}(\mathbf{x}_1, \cdots, \mathbf{x}_{p+q})$. At first, the term $(l_p \wedge^* l_q) \circ \psi_{p+q}(\mathbf{x}_1, \cdots, \mathbf{x}_{p+q})$ is given by:

$$\sum_{\sigma \in S_{p+q}} \frac{\varepsilon(\sigma)}{(p+q)!} l_p \circ \psi_p \left(\mathbf{x}_{\sigma(1)}, \cdots, \mathbf{x}_{\sigma(p)}\right) l_q \circ \psi_q \left(\mathbf{x}_{\sigma(p+1)}, \cdots, \mathbf{x}_{\sigma(p+q)}\right), \quad [3.51]$$

whereas the term $(l_q \wedge^* l_p) \circ \psi_{p+q}(\mathbf{x}_1, \cdots, \mathbf{x}_{p+q})$ is given by the formula:

$$\sum_{\sigma \in S_{p+q}} \frac{\varepsilon(\sigma)}{(p+q)!} l_p \circ \psi_p \left(\mathbf{x}_{\sigma(q+1)}, \cdots, \mathbf{x}_{\sigma(q+p)}\right) l_q \circ \psi_q \left(\mathbf{x}_{\sigma(1)}, \cdots, \mathbf{x}_{\sigma(q)}\right). \quad [3.52]$$

Now, let us consider the permutation $\alpha \in S_{p+q}$ that maps $1 \to q+1, \cdots p \to p+q$, then $p+1 \to 1, \cdots p+q \to q$. It is clear that we have:

$$[l_p \wedge^* l_q] \circ \psi_{p+q}(\mathbf{x}_1, \cdots, \mathbf{x}_{p+q}) = [l_q \wedge^* l_p] \circ \psi_{p+q}\left(\mathbf{x}_{\alpha(1)}, \cdots, \mathbf{x}_{\alpha(p+q)}\right). \quad [3.53]$$

According to the calculations above (see formula [3.50] applied to the right-hand side of the previous equality), we have thus:

$$[l_p \wedge^* l_q] \circ \psi_{p+q}(\mathbf{x}_1, \cdots, \mathbf{x}_{p+q}) = \varepsilon(\alpha) [l_q \wedge^* l_p] \circ \psi_{p+q}(\mathbf{x}_1, \cdots, \mathbf{x}_{p+q}), \quad [3.54]$$

and the same can be applied to any family $\mathbf{x}_1, \cdots, \mathbf{x}_{p+q}$ of $\boldsymbol{E}^{p+q}$. As a result, we deduce the equality:

$$l_p \wedge^* l_q = \varepsilon(\alpha) l_q \wedge^* l_p. \quad [3.55]$$

Therefore, it is a matter of calculating $\varepsilon(\alpha)$. For this purpose, it is necessary to find a decomposition of α into a transposition sequence. This is easily achieved. In order to move element 1 towards element $q+1$ it suffices to transpose 1 with 2, then to transpose 2 with 3 and so forth up to transposing q with $q+1$: q transpositions are required for exchanging 1 and $q+1$, and maintaining the other elements constant. Idem for exchanging 2 with $q+2$, up to p with $q+p$. Therefore, the permutation α can be decomposed by a product of qp transpositions. Hence, we deduce:

$$\varepsilon(\alpha) = (-1)^{pq} \Rightarrow l_p \wedge^* l_q = (-1)^{pq} l_q \wedge^* l_p. \qquad [3.56]$$

4) The p-linearity is immediate. As for the skew symmetry, it suffices to consider a transposition $\tau \in S_p$. Let us suppose that this transposition exchanges the elements ϕ_r, ϕ_s within the calculation of:

$$\phi_1 \wedge^* \cdots \wedge^* \phi_p. \qquad [3.57]$$

Then, in order to "move back" the element ϕ_r towards ϕ_s (with $r < s$), it is necessary to perform $r-s-1$ "exchanges" adjacent linear form functions ϕ_i, ϕ_{i+1}. Each time that we exchange adjacent linear forms, we multiply by -1. Once we obtain ϕ_s beside ϕ_r, we perform an exchange (of adjacent forms), and then we perform $r-s-1$ exchanges in order to "move back" the form ϕ_r to position s. In total, exactly $2s-2r-1$ exchanges of adjacent forms are performed; therefore, in order to exchange two elements, we multiply the term by -1. The map:

$$(\phi_1, \cdots, \phi_p) \mapsto \phi_1 \wedge^* \cdots \wedge^* \phi_p \qquad [3.58]$$

is indeed skew-symmetric.

3.2.3. *Characterization of linear independent families*

PROPOSITION 3.7.– Let $L : \boldsymbol{E} \mapsto \boldsymbol{E}^*$ be an injection from $\boldsymbol{E}$ to $\boldsymbol{E}^*$ that maps the algebraic basis $\mathbf{e}_i, i \in I$ over to the associated dual basis $\phi_i^*, i \in I$ (see proposition 1.5). Then, there exists a unique linear map $\mathcal{L} : \Lambda^p\boldsymbol{E} \mapsto (\Lambda^p\boldsymbol{E})^*$ such that:

$$\forall \mathbf{x}_1, \cdots, \mathbf{x}_p,\ \ \mathcal{L}(\mathbf{x}_1 \wedge \cdots \wedge \mathbf{x}_p) = L(\mathbf{x}_1) \wedge^* \cdots \wedge^* L(\mathbf{x}_p). \qquad [3.59]$$

PROOF.– In virtue of linearity of the map L, and since the symbol $\wedge^*$ enables the definition of an alternating p-linear symbol with values in $\mathcal{A}(\boldsymbol{E}^p, \mathbb{R})$, it follows immediately that the map:

$$\mathbf{x}_1, \cdots, \mathbf{x}_p \mapsto L(\mathbf{x}_1) \wedge^* \cdots \wedge^* L(\mathbf{x}_p) \qquad [3.60]$$

is a p-linear skew-symmetric map over $\boldsymbol{E}^p$ with values in $(\Lambda^p\boldsymbol{E})^*$. According to the construction theorem 2.1, there exists a unique linear map $\mathcal{L} : \Lambda^p\boldsymbol{E} \mapsto (\Lambda^p\boldsymbol{E})^*$ such that:

$$\mathcal{L}(\mathbf{x}_1 \wedge \cdots \wedge \mathbf{x}_p) = L(\mathbf{x}_1) \wedge^* \cdots \wedge^* L(\mathbf{x}_p), \qquad [3.61]$$

which concludes our demonstration.

Let us consider now a meaningful result:

PROPOSITION 3.8.– Let $\mathbf{x}_1, \cdots, \mathbf{x}_p$ be a linear independent family of $\boldsymbol{E}$. Then, we have $\mathbf{x}_1 \wedge \cdots \wedge \mathbf{x}_p \neq \mathbf{0}$.

PROOF.– Since the family $\mathbf{x}_1, \cdots, \mathbf{x}_p$ is independent, it constitutes a basis of a linear space $\boldsymbol{F} := \langle \mathbf{x}_1, \cdots, \mathbf{x}_p \rangle$. According to proposition 1.5, we can consider the dual basis $\phi_j^*, j \in [1, p]$ and a bijective linear map $Q : \boldsymbol{F} \mapsto \boldsymbol{F}^*$ such that $Q(\mathbf{x}_i) = \phi_i^*$. The alternating p-linear form defined by:

$$\phi_1^* \wedge^* \cdots \wedge^* \phi_p^* := Q(\mathbf{x}_1) \wedge^* \cdots \wedge^* Q(\mathbf{x}_p) \qquad [3.62]$$

is non (identically) zero. In fact, we can easily calculate:

$$\left[\phi_1^* \wedge^* \cdots \wedge^* \phi_p^*\right](\mathbf{x}_1, \cdots, \mathbf{x}_p) = \frac{1}{p!} \sum_{\sigma \in S_p} \varepsilon(\sigma)\, \phi_{\sigma(1)}^*(\mathbf{x}_1) \cdots \phi_{\sigma(p)}^*(\mathbf{x}_p). \qquad [3.63]$$

We know that $\phi_k^*(\mathbf{x}_l) = \delta(k, l)$. Therefore, we have:

$$\left[\phi_1^* \wedge^* \cdots \wedge^* \phi_p^*\right](\mathbf{x}_1, \cdots, \mathbf{x}_p) = \frac{1}{p!} \sum_{\sigma \in S_p} \varepsilon(\sigma)\, \delta(1, \sigma(1)) \cdots \delta(p, \sigma(p)). \qquad [3.64]$$

This summation over all the permutations has only one single term that is non-zero, which is for $\sigma = Id$. Hence, we deduce:

$$\left[\phi_1^* \wedge^* \cdots \wedge^* \phi_p^*\right](\mathbf{x}_1, \cdots, \mathbf{x}_p) = \frac{1}{p!} \neq 0. \qquad [3.65]$$

Now, according to the previous proposition 3.7, there exists a (unique) linear map $\mathcal{Q} : \Lambda^p \boldsymbol{F} \mapsto (\Lambda^p \boldsymbol{F})^*$ such that:

$$\left[\phi_1^* \wedge^* \cdots \wedge^* \phi_p^*\right](\mathbf{x}_1, \cdots, \mathbf{x}_p) = \mathcal{Q}(\mathbf{x}_1 \wedge \cdots \wedge \mathbf{x}_p)\left[\mathbf{x}_1, \cdots, \mathbf{x}_p\right]. \qquad [3.66]$$

If we had $\mathbf{x}_1 \wedge \cdots \wedge \mathbf{x}_p = \mathbf{0}$, since $\mathcal{Q}$ is linear, we would have:

$$\mathcal{Q}(\mathbf{x}_1 \wedge \cdots \wedge \mathbf{x}_p) = 0 \qquad [3.67]$$

being the case of an alternating multi-linear map over $\boldsymbol{E}^p$ and thus we would have:

$$\mathcal{Q}(\mathbf{x}_1 \wedge \cdots \wedge \mathbf{x}_p)\left[\mathbf{x}_1, \cdots, \mathbf{x}_p\right] = 0, \qquad [3.68]$$

which contradicts what we have just demonstrated.

Consider now the reciprocal proposition (enunciated in a counterposed way):

PROPOSITION 3.9.– Let $\mathbf{x}_1, \cdots, \mathbf{x}_p$ be a linearly dependent family. Then, we have $\mathbf{x}_1 \wedge \cdots \wedge \mathbf{x}_p = \mathbf{0}$.

PROOF.– Since the family is dependent, there exists a vector which is a linear combination of other vectors. Without loss of generality, we can suppose it is $\mathbf{x}_p$ that can be written as:

$$\mathbf{x}_p = \sum_{k=1}^{k=p-1} \chi_k \mathbf{x}_k. \quad [3.69]$$

In virtue of p-linearity, we calculate then:

$$\mathbf{x}_1 \wedge \cdots \wedge \mathbf{x}_p = \sum_{k=1}^{k=p-1} \chi_k \mathbf{x}_1 \wedge \cdots \wedge \mathbf{x}_{p-1} \wedge \mathbf{x}_k \quad [3.70]$$

Now, within the summation above, all the exterior products have at least two equal terms, which ensures their nullity due to the skew symmetry of exterior products $\wedge$, hence the proof of the proposition.

Finally, we have the theorem:

THEOREM 3.2.– A family of p vectors $\mathbf{x}_1, \cdots, \mathbf{x}_p$ is linearly independent if and only if $\mathbf{x}_1 \wedge \cdots \wedge \mathbf{x}_p \neq \mathbf{0}$.

4

Bases of Exterior Algebras

In this chapter, we shall tackle the following fundamental question:

QUESTION 4.1.– Is it possible to construct bases for spaces $\Lambda^p \boldsymbol{E}$ starting from known bases in $\boldsymbol{E}$? Similarly, is it then possible to characterize the space $\Lambda^p \boldsymbol{E}^*$?

Fortunately, the answer to this question is affirmative. A remarkable characteristic of such a construction is the fact that it requires, in the first place, a definition of the exterior product. In other words, without the exterior product it would not necessarily be possible to illustrate the bases of $\Lambda^p \boldsymbol{E}$ (or of $\Lambda^p \boldsymbol{E}^*$) in a constructive way, which would make the usage of such spaces irrelevant.

4.1. Construction of algebraic bases

4.1.1. *Generating a set of exterior algebras*

PROPOSITION 4.1.– Let $\langle \psi_p (\boldsymbol{E}^p) \rangle$ be a linear space generated by a set of images of elements from $\boldsymbol{E}^p$ mapped by ψ_p. In that case, we have:

$$\Lambda^p \boldsymbol{E} = \langle \psi_p (\boldsymbol{E}^p) \rangle. \qquad [4.1]$$

PROOF.– Let us denote by $\boldsymbol{G}$ the linear space generated by $\psi_p (\boldsymbol{E}^p)$. Suppose that we have $\Lambda^p \boldsymbol{E} = \boldsymbol{G} \oplus \boldsymbol{H}$ with $\boldsymbol{H}$ a complementary space with respect to $\boldsymbol{G}$ in $\Lambda^p \boldsymbol{E}$ (the existence of a complementary space for any linear subspace is guaranteed by the incomplete basis theorem, see corollary 1.1). Let $\boldsymbol{X}$ be a linear space. It is clear that the set $\mathcal{L} (\Lambda^p \boldsymbol{E}, \boldsymbol{X})$ is isomorphic to the product $\mathcal{L} (\boldsymbol{G}, \boldsymbol{X}) \times \mathcal{L} (\boldsymbol{H}, \boldsymbol{X})$. In other words, in order to define any linear map from $\Lambda^p \boldsymbol{E}$ into $\boldsymbol{X}$, it is necessary and sufficient to specify how this map operates respectively over $\boldsymbol{G}$ and over $\boldsymbol{H}$. Let thus $l_1 := (Id, \lambda_1) \in \mathcal{L}(\boldsymbol{G},\boldsymbol{X}) \times \mathcal{L}(\boldsymbol{H},\boldsymbol{X})$ and $l_2 := (Id, \lambda_2) \in \mathcal{L}(\boldsymbol{G},\boldsymbol{X}) \times \mathcal{L}(\boldsymbol{H},\boldsymbol{X})$ be two linear maps over $\boldsymbol{F}$ with values in $\boldsymbol{X}$. Then, the maps:

$$\gamma_p^1 := l_1 \circ \psi_p, \quad \gamma_p^2 := l_2 \circ \psi_p, \tag{4.2}$$

are two alternating p-linear maps that are coincident over $\boldsymbol{E}^p$. If it would be possible to find $\lambda_1 \neq \lambda_2 \in \mathcal{L}(\boldsymbol{H},\boldsymbol{X})$, then it would be possible to find two different linear maps l_1, l_2 from $\Lambda^p \boldsymbol{E}$ into $\boldsymbol{X}$ such that for $\gamma_p := \gamma_p^1 = \gamma_p^2$ an alternating p-linear map over $\boldsymbol{E}^p$, we would have:

$$\gamma_p = l_1 \circ \psi_p = l_2 \circ \psi_p, \tag{4.3}$$

which would contradict the definition of $\Lambda^p \boldsymbol{E}$ being an exterior algebra with constructor ψ_p. Therefore, we have, for any linear space $\boldsymbol{X}$, that $\mathcal{L}(\boldsymbol{H},\boldsymbol{X}) = \{0\}$. This implies that $\boldsymbol{H} = \{\mathbf{0}\}$.

COROLLARY 4.1.– Let $\boldsymbol{E}$ be a linear space having $\mathbf{e}_i, i \in I$ as an algebraic basis, such that I is equipped with a totally ordered structure. In that case, the family of all the elements of the form:

$$\psi_p\left(\mathbf{e}_{i_1}, \cdots, \mathbf{e}_{i_p}\right), \; i_1 \leq \cdots \leq i_p \tag{4.4}$$

is a generating set of $\Lambda^p \boldsymbol{E}$. If the subset of strictly ordered multi-indexes is non-empty, then the family of

$$\psi_p\left(\mathbf{e}_{i_1}, \cdots, \mathbf{e}_{i_p}\right), \; i_1 < \cdots < i_p \tag{4.5}$$

is also a generating set of the space $\Lambda^p \boldsymbol{E}$.

PROOF.– It suffices to note that, in virtue of multi-linearity, all the elements of the form $\psi_p(\mathbf{x}_1, \cdots, \mathbf{x}_p)$ can be written as a finite linear combination of elements of the form:

$$\psi_p\left(\mathbf{e}_{j_1}, \cdots, \mathbf{e}_{j_p}\right), (j_1, \cdots, j_p) \in I^p \tag{4.6}$$

now, since I is totally ordered, by performing a permutation of the indexes, it is possible to obtain an ordered multi-index (see definition 1.9) $i_1 \leq \cdots \leq i_p$ and a permutation $\sigma \in S_p$ such that we have:

$$\forall k \in [1,p], \; i_k = j_{\sigma(k)}. \tag{4.7}$$

By employing the fact that ψ_p is an alternating p-linear map, we have:

$$\psi_p\left(\mathbf{e}_{j_1},\cdots,\mathbf{e}_{j_p}\right)=\varepsilon(\sigma)\,\psi_p\left(\mathbf{e}_{i_1},\cdots,\mathbf{e}_{i_p}\right), \qquad [4.8]$$

and we reach the result: any linear combination (with finite support) of elements $\psi_p(\mathbf{x}_1,\cdots,\mathbf{x}_p)$ can be written as a linear combination with finite support of $\psi_p\left(\mathbf{e}_{i_1},\cdots,\mathbf{e}_{i_p}\right)$ with $i_1\leq\cdots\leq i_p$. Now, in virtue of skew symmetry, among the elements of the form $\psi_p\left(\mathbf{e}_{i_1},\cdots,\mathbf{e}_{i_p}\right)$, $i_1\leq\cdots\leq i_p$ we can remove all the elements (which are necessarily null) with two equal successive indexes. Therefore, the remaining elements (since, according to the corollary's hypothesis, such a set is non-empty) are constituted only by the family of $\psi_p\left(\mathbf{e}_{i_1},\cdots,\mathbf{e}_{i_p}\right)$ with strictly ordered multi-indexes $i_1<\cdots<i_p$.

COROLLARY 4.2.– Let $\boldsymbol{E}$ be a linear space of dimension n and p an integer. If $p>n$ then $\boldsymbol{F}=\{\mathbf{0}\}$ and $\psi_p=0$. In other words, for any space $\boldsymbol{E}$ with a finite dimension n, we have $\Lambda^p\boldsymbol{E}=\{\mathbf{0}\}$ as soon as $p>n$.

PROOF.– If $p>n$ then every set of the form $\mathbf{e}_{i_1},\cdots,\mathbf{e}_{i_p}$ with $1\leq i_1\leq\cdots\leq i_p\leq n$ necessarily contains two equal elements, and thus $\psi_p\left(\mathbf{e}_{i_1},\cdots,\mathbf{e}_{i_p}\right)$ is zero. Therefore, $\boldsymbol{F}$, which is generated by this family, is a null linear space and ψ_p is a null map.

4.1.2. *Linearly independent family of exterior algebras*

THEOREM 4.1.– Let $\boldsymbol{E}$ be a linear space (such that $\Lambda^p\boldsymbol{E}\neq\{\mathbf{0}\}$) with an algebraic basis $\mathbf{e}_i, i\in I$ where I is a totally ordered set. In that case, the family of

$$\mathbf{e}_{i_1}\wedge\cdots\wedge\mathbf{e}_{i_p},\quad i_1<\cdots<i_p\in\mathcal{I}_o(p) \qquad [4.9]$$

is a linearly independent family of $\Lambda^p\boldsymbol{E}$.

PROOF.– Consider a linear combination (with finite support) that equals zero:

$$\sum_{i_1<\cdots<i_p}\alpha_{i_1,\cdots,i_p}\mathbf{e}_{i_1}\wedge\cdots\wedge\mathbf{e}_{i_p}=\mathbf{0}. \qquad [4.10]$$

Since the family of $\alpha_{i_1,\cdots,i_p}$ has a finite support, this means that there exists a finite subset of indexes $J\subset I$ such that in the case where for $\alpha_{i_1,\cdots,i_p}$, there exists an index i_k which is not in J, the coefficient $\alpha_{i_1,\cdots,i_p}$ is equal to zero. In other words, since the family of $\alpha_{i_1,\cdots,i_p}$ has a finite support, we can always suppose that the values of indexes i_k belong to a finite ordered set $J=[1,n]$. In that case, we have that:

$$\sum_{1\leq i_1<\cdots<i_p\leq n}\alpha_{i_1,\cdots,i_p}\mathbf{e}_{i_1}\wedge\cdots\wedge\mathbf{e}_{i_p}=\mathbf{0}. \qquad [4.11]$$

Consider $\mathbf{e}_{j_1} \wedge \cdots \wedge \mathbf{e}_{j_{n-p}}$, where $j_1, \cdots, j_{n-p}$ is a strictly ordered $n-p$ index. We have immediately that:

$$\forall (j_1, \cdots, j_{n-p}) \in \mathcal{I}_o(n-p),$$

$$\left(\sum_{1 \le i_1 < \cdots < i_p \le n} \alpha_{i_1,\cdots,i_p} \mathbf{e}_{i_1} \wedge \cdots \wedge \mathbf{e}_{i_p} \right) \wedge \left(\mathbf{e}_{j_1} \wedge \cdots \wedge \mathbf{e}_{j_{n-p}} \right) = \mathbf{0} \quad [4.12]$$

by developing, in virtue of linearity, we thus deduce that we have:

$$\forall (j_1, \cdots, j_{n-p}) \in \mathcal{I}_o(n-p),$$

$$\sum_{1 \le i_1 < \cdots < i_p \le n} \alpha_{i_1,\cdots,i_p} \mathbf{e}_{i_1} \wedge \cdots \wedge \mathbf{e}_{i_p} \wedge \mathbf{e}_{j_1} \wedge \cdots \wedge \mathbf{e}_{j_{n-p}} = \mathbf{0}, \quad [4.13]$$

let us note now that the element $\mathbf{e}_{i_1} \wedge \cdots \wedge \mathbf{e}_{i_p} \wedge \mathbf{e}_{j_1} \wedge \cdots \wedge \mathbf{e}_{j_{n-p}}$ is an exterior product of n vectors. Only two cases are possible:

1) Either there exist two equal indexes, or in other words, $\{j_1, \cdots, j_{n-p}\} \cap \{i_1, \cdots i_p\} \neq \varnothing$. In that case, it is immediate that we have:

$$\mathbf{e}_{i_1} \wedge \cdots \wedge \mathbf{e}_{i_p} \wedge \mathbf{e}_{j_1} \wedge \cdots \wedge \mathbf{e}_{j_{n-p}} = \mathbf{0} \quad [4.14]$$

2) or $\{j_1, \cdots, j_{n-p}\} \cap \{i_1, \cdots i_p\} = \varnothing$. This can happen in only one situation, i.e. when the set $\{j_1, \cdots, j_{n-p}\}$ is the (ordered) complementary of $\{i_1, \cdots i_p\}$ (then denoted by $C_I^o \{i_1, \cdots i_p\}$, which is thus an ordered $n-p$-index). In such a case, it is clear that we have:

$$\mathbf{e}_{i_1} \wedge \cdots \wedge \mathbf{e}_{i_p} \wedge \mathbf{e}_{j_1} \wedge \cdots \wedge \mathbf{e}_{j_{n-p}} \neq \mathbf{0}, \quad [4.15]$$

which is due to the fact that the family $\mathbf{e}_1, \cdots, \mathbf{e}_n$ is linearly independent (because it is extracted from a basis, see corollary 1.1) and therefore we can apply proposition 3.8.

We thus reach the conclusion that:

$$\sum_{1 \le i_1 < \cdots < i_p \le n} \alpha_{i_1,\cdots,i_p} \mathbf{e}_{i_1} \wedge \cdots \wedge \mathbf{e}_{i_p} \wedge \mathbf{e}_{j_1} \wedge \cdots \wedge \mathbf{e}_{j_{n-p}} =$$

$$\alpha_{C_I^o\{j_1,\cdots,j_{n-p}\}} \varepsilon(\sigma) \mathbf{e}_1 \wedge \cdots \wedge \mathbf{e}_n, \quad [4.16]$$

where σ is a permutation of $[1, n]$ that maps $(1, \cdots, n)$ into $(i_1, \cdots, i_p, j_1, \cdots j_{n-p})$. As a result, we have:

$$\forall j_1, \cdots j_{n-p} \in \mathcal{I}_o(n-p), \quad \alpha_{C_I^o\{j_1,\cdots,j_{n-p}\}} = 0. \quad [4.17]$$

However, it is clear that if $j_1, \cdots j_{n-p}$ spans $\mathcal{I}_o(n-p)$, then its (ordered) complementary spans $\mathcal{I}_o(p)$. Therefore, each coefficient among $\alpha_{i_1,\cdots,i_p}$ with $(i_1, \cdots, i_p) \in \mathcal{I}_o(p)$ is zero. Hence, the family is indeed a linearly independent family.

4.1.3. *Finite dimensional Grassmann's algebra*

We have thus reached the following theorem:

THEOREM 4.2.– Let $\boldsymbol{E}$ be a linear space that admits $\mathbf{e}_i, i \in I$, where I is a totally ordered set. Then, if it is non-empty, the family of:

$$\mathbf{e}_{i_1} \wedge \cdots \wedge \mathbf{e}_{i_p}, \quad \mathbf{i} := (i_1, \cdots, i_p) \in \mathcal{I}_o(p) \qquad [4.18]$$

is an algebraic basis of $\Lambda^p \boldsymbol{E}$.

PROOF.– This theorem is a corollary of the two previous properties: such a family is both linearly independent and a generating set in $\Lambda^p \boldsymbol{E}$.

We have the following corollary:

COROLLARY 4.3.– Let $\boldsymbol{E}$ be a linear space with a finite dimension n. In that case, $\Lambda^p \boldsymbol{E}$ has a finite dimension and we have $\dim(\Lambda^p \boldsymbol{E}) = \binom{n}{p}$, with the common convention that $\binom{n}{p} = 0$ for $n < p$.

PROOF.– Since every finite dimensional linear space admits a totally ordered basis $\mathbf{e}_i, \in [1, n]$, we have $\Lambda^p \boldsymbol{E} \neq \{\mathbf{0}\}$ if and only if the set of strictly ordered p-indexes is non-empty. In accordance with remark 1.3, this is possible only if $p \leq n$. In such a case, there are exactly $\binom{n}{p}$ p-indexes to be constructed, which yields the dimension of the linear space.

A result of this theorem is that Grassmann's algebra takes a particular form when it is constructed over a linear space with a finite dimension n: all the sets $\Lambda^p \boldsymbol{E}$ are null for $p > n$. Therefore, it is clear that we can state the following corollary:

COROLLARY 4.4.– Let $\boldsymbol{E}$ be a linear space with a finite dimension n. In that case, we have:

$$\mathcal{G}(\boldsymbol{E}) = \mathbb{R} \times \boldsymbol{E} \times \Lambda^2 \boldsymbol{E} \times \cdots \times \Lambda^n \boldsymbol{E} \times \{0\} \times \cdots \cong \mathbb{R} \times \boldsymbol{E} \times \Lambda^2 \boldsymbol{E} \times \cdots \times \Lambda^n \boldsymbol{E}, \qquad [4.19]$$

where the symbol $\cong$ represents the existence of an isomorphism (which is evident here) between the sets. The dimension of Grassmann's algebra is then equal to:

$$\dim(\mathcal{G}) = 1 + \binom{n}{1} + \cdots + \binom{n}{n-1} + \binom{n}{n} = 2^n. \qquad [4.20]$$

4.2. Identification of the dual of $\Lambda^p E$

4.2.1. *Identification of alternating* 2*-linear forms over* $\mathbb{R}^3$

In this section, we consider an alternating 2-linear form over $\mathbb{R}^3$. It is clear that such a form is defined as soon as we know the expression of $a(\mathbf{e}_i, \mathbf{e}_j)$ for $1 \leq i < j \leq 3$. We shall thus set $a_{ij} := a(\mathbf{e}_i, \mathbf{e}_j)$. Let us consider the dual basis ϕ_i^* associated with the basis $\mathbf{e}_i, i \in [1, 3]$ and let us consider the bilinear form:

$$b(\cdot,\cdot) = \sum_{i<j} a_{ij}\phi_j^* \wedge^* \phi_i^* (\cdot,\cdot). \qquad [4.21]$$

We can calculate the value of this function for all the pairs $\mathbf{e}_k, \mathbf{e}_l$:

$$\begin{aligned} b(\mathbf{e}_k, \mathbf{e}_l) &= \sum_{i<j} a_{ij}\phi_j^* \wedge^* \phi_i^* (\mathbf{e}_k, \mathbf{e}_l) \\ &= \sum_{i<j} a_{ij} \frac{1}{2!} \left[\phi_i^*(\mathbf{e}_k)\phi_j(\mathbf{e}_l) - \phi_j^*(\mathbf{e}_k)\phi_i(\mathbf{e}_l)\right], \end{aligned} \qquad [4.22]$$

which can be rewritten as:

$$b(\mathbf{e}_k, \mathbf{e}_l) = \sum_{i<j} a_{ij} \frac{1}{2!} \left[\delta(i,k)\delta(j,l) - \delta(j,k)\delta(i,l)\right]. \qquad [4.23]$$

It is immediate that b is skew-symmetric, and thus we can limit ourselves to $k < l$. In such a case, the term $\delta(j,k)\delta(i,l)$ is always null, since $i < j$. Therefore, it remains simply:

$$\forall k < l, \;\; b(\mathbf{e}_k, \mathbf{e}_l) = \frac{1}{2}\sum_{i<j} a_{ij}\delta(i,k)\delta(j,l) = \frac{1}{2}a_{kl}. \qquad [4.24]$$

Thus, in the end, we have $a(\cdot,\cdot) = 2b(\cdot,\cdot)$ for all the vectors. This means that any skew-symmetric bilinear form $a(\cdot,\cdot)$ can be decomposed as an exterior product $\wedge^*$ of maps of the form $\phi_i^* \wedge^* \phi_j^*, 1 \leq i < j \leq 3$: this family is thus a generating set of the space of skew-symmetric 2-linear forms over $\mathbb{R}^3$. Another result arises from the fact that this family is also a linearly independent family, since if we suppose that:

$$\sum_{i<j} a_{ij}\phi_i^* \wedge^* \phi_j^* = 0 \qquad [4.25]$$

then, by performing a calculation for all the arguments of the form $\mathbf{e}_k, \mathbf{e}_l$ (with $k < l$), we obtain directly $a_{kl} = 0$: hence the fact that the family is indeed linearly independent. Finally, we reach the following lemma:

LEMMA 4.1.– In $\mathbb{R}^3$, the linear space of alternating 2-linear forms admits the family of $\phi_i^* \wedge^* \phi_j^*, i < j$ as a basis.

The following question arises:

QUESTION 4.2.– Is it possible to generate a basis of the space of alternating p-linear forms over $\boldsymbol{E}^p$ from the exterior product $\wedge^*$ of linear forms?

4.2.2. *Basis of alternating p-linear forms over $\boldsymbol{E}^p$*

The answer to the question above will prove to be positive, although on the condition of working within a finite dimensional space $\boldsymbol{E}$.

THEOREM 4.3.– Let $\phi_i^*, i \in [1, n]$ be a basis of the space $\boldsymbol{E}^*$. Let $\mathcal{A}(\boldsymbol{E}^p, \mathbb{R})$ be a space of alternating p-linear forms over $\boldsymbol{E}^p$. In that case, the family of $\phi_{i_1}^* \wedge^* \cdots \wedge^* \phi_{i_p}^*$ with $\mathbf{i} = (i_1, \cdots, i_p) \in \mathcal{I}_o(p)$ is a basis of $\mathcal{A}(\boldsymbol{E}^p, \mathbb{R})$.

PROOF.– In virtue of the isomorphism between $\boldsymbol{E}$ and its dual $\boldsymbol{E}^*$, we can always consider the family $\phi_i^*, i \in [1, n]$ as the dual basis of a basis $\mathbf{e}_i, i \in [1, n]$ of the space $\boldsymbol{E}$. By skew symmetry and p-linearity, in order to characterize a skew-symmetric p-linear map over $\boldsymbol{E}^p$, it is necessary and sufficient to consider a set of values:

$$a(\mathbf{e}_{i_1}, \cdots, \mathbf{e}_{i_p}), \quad \mathbf{i} := (i_1, \cdots, i_p) \in \mathcal{I}_o(p). \qquad [4.26]$$

Two skew-symmetric p-linear maps with identical values over this family of (ordered) vectors are necessarily equal. Let us now consider a skew-symmetric p-linear form given by:

$$b := \sum_{\mathbf{i} \in \mathcal{I}_o(p)} a(\mathbf{e}_{i_1}, \cdots, \mathbf{e}_{i_p}) \, \phi_{i_1}^* \wedge^* \cdots \wedge^* \phi_{i_p}^*, \qquad [4.27]$$

we calculate the value of this form for the lists of $\boldsymbol{E}^p$ of the form $(\mathbf{e}_{j_1}, \cdots, \mathbf{e}_{j_p})$ with $\mathbf{j} \in \mathcal{I}_o(p)$ a strictly ordered p-index. We then have:

$$b(\mathbf{e}_{j_1}, \cdots, \mathbf{e}_{j_p}) = \sum_{\mathbf{i} \in \mathcal{I}_o(p)} a(\mathbf{e}_{i_1}, \cdots, \mathbf{e}_{i_p}) \frac{1}{p!} \sum_{\sigma \in S_p} \varepsilon(\sigma) \, \phi_{i_{\sigma(1)}}^*(\mathbf{e}_{j_1}) \cdots \phi_{i_{\sigma(p)}}^*(\mathbf{e}_{j_p}). \qquad [4.28]$$

This can be rewritten as:

$$b\left(\mathbf{e}_{j_1},\cdots,\mathbf{e}_{j_p}\right)=\sum_{\mathbf{i}\in\mathcal{I}_o(p)} a\left(\mathbf{e}_{i_1},\cdots,\mathbf{e}_{i_p}\right)\frac{1}{p!}$$

$$\sum_{\sigma\in S_p}\varepsilon(\sigma)\,\delta\left(i_{\sigma(1)},j_1\right)\cdots\delta\left(i_{\sigma(p)},j_p\right) \qquad [4.29]$$

now, since the p-index $\mathbf{j}$ is strictly ordered, a necessary condition in order to have a non-null term of the form $\delta\left(i_{\sigma(1)},j_1\right)\cdots\delta\left(i_{\sigma(p)},j_p\right)$ is that the p-index $i_{\sigma(1)},\cdots,i_{\sigma(p)}$ also has to be strictly ordered (because, in order to have non-nullity, it is necessary and sufficient to have $i_{\sigma_k}=j_k$ for all the integers $k\in[1,p]$). Since $i_1,\cdots,i_p$ is already strictly ordered, this implies that the only permutation which can preserve the order is the identity permutation. In virtue of the order over the p-index $\mathbf{j}$, we have thus:

$$b\left(\mathbf{e}_{j_1},\cdots,\mathbf{e}_{j_p}\right)=\sum_{\mathbf{i}\in\mathcal{I}_o(p)} a\left(\mathbf{e}_{i_1},\cdots,\mathbf{e}_{i_p}\right)\frac{1}{p!}\delta\left(i_1,j_1\right)\cdots\delta\left(i_p,j_p\right), \qquad [4.30]$$

which is non-null if and only if $\mathbf{i}=\mathbf{j}$. Hence, the result:

$$b\left(\mathbf{e}_{j_1},\cdots,\mathbf{e}_{j_p}\right)=\frac{1}{p!}a\left(\mathbf{e}_{j_1},\cdots,\mathbf{e}_{j_p}\right). \qquad [4.31]$$

Therefore, we deduce that $p!b=a$, over all the $\left(\mathbf{e}_{i_1},\cdots,\mathbf{e}_{i_p}\right)$ with $\mathbf{i}\in\mathcal{I}_o$ and thus, by p-linearity and by skew symmetry, this equality is true for all the elements of $\boldsymbol{E}^p$. The family of $\phi^*_{i_1}\wedge^*\cdots\wedge^*\phi^*_{i_p}$ with $\mathbf{i}=(i_1,\cdots,i_p)\in\mathcal{I}_o(p)$ is a generating set of $\mathcal{A}(\boldsymbol{E}^p,\mathbb{R})$. However, by means of a similar calculation, it can be shown that if we have:

$$\sum_{\mathbf{i}\in\mathcal{I}_o(p)} a_{i_1,\cdots,i_p}\phi^*_{i_1}\wedge^*\cdots\wedge^*\phi^*_{i_p}=0 \qquad [4.32]$$

then, by calculating the form over the vector $\left(\mathbf{e}_{j_1},\cdots,\mathbf{e}_{j_p}\right)$ with $\mathbf{j}\in\mathcal{I}_o(p)$, we have immediately $a_{j_1,\cdots,j_p}=0$ and thus the family is also a linearly independent family, which concludes the demonstration.

REMARK 4.1.– The key point of the demonstration lies in the fact that any basis of $\boldsymbol{E}^*$ can be considered as a dual basis of a basis of $\boldsymbol{E}$, which can happen only in the case of finite dimension. There is no possibility of such a theorem being true in the case of infinite dimension. ■

4.2.3. *Skew-symmetric form and dual space*

THEOREM 4.4.– The space $\mathcal{A}(\boldsymbol{E}^p,\mathbb{R})$ is isomorphic to the dual space $\left(\Lambda^p\boldsymbol{E}\right)^*$.

PROOF.– An alternating p-linear form is by definition a skew-symmetric p-linear map with values in $\mathbb{R}$. In virtue of the fundamental theorem of construction of exterior algebra, for all $a \in \mathcal{A}(\boldsymbol{E}^p, \mathbb{R})$ there exists a unique map $l(a) : \Lambda^p \boldsymbol{E} \mapsto \mathbb{R}$ (and thus there exists a linear form over $\Lambda^p \boldsymbol{E}$, i.e. an element of $(\Lambda^p \boldsymbol{E})^*$) such that we have:

$$\forall \mathbf{x}_1, \cdots, \mathbf{x}_p \in \boldsymbol{E}^p, \quad a(\mathbf{x}_1, \cdots, \mathbf{x}_p) = l(a)(\mathbf{x}_1 \wedge \cdots \wedge \mathbf{x}_p). \qquad [4.33]$$

Therefore, it is possible to define, in a univocal way, a map $l(\cdot) : \mathcal{A}(\boldsymbol{E}^p, \mathbb{R}) \mapsto (\Lambda^p \boldsymbol{E})^*$. We shall show that this map is actually an isomorphism. It is clear that if $a_1, a_2 \in \mathcal{A}(\boldsymbol{E}^p, \mathbb{R})$ and if $\lambda \in \mathbb{R}$, then we have:

$$[l(a_1) + \lambda l(a_2)](\mathbf{x}_1 \wedge \cdots \wedge \mathbf{x}_p) = a_1(\mathbf{x}_1, \cdots, \mathbf{x}_p) + \lambda a_2(\mathbf{x}_1, \cdots, \mathbf{x}_p). \qquad [4.34]$$

However, by definition, the right-hand side of this equality can be written as:

$$\begin{aligned} a_1(\mathbf{x}_1, \cdots, \mathbf{x}_p) + \lambda a_2(\mathbf{x}_1, \cdots, \mathbf{x}_p) &= (a_1 + \lambda a_2)(\mathbf{x}_1, \cdots, \mathbf{x}_p) \\ &= l(a_1 + \lambda a_2)(\mathbf{x}_1 \wedge \cdots \wedge \mathbf{x}_p), \end{aligned} \qquad [4.35]$$

and we immediately deduce thereof that $l(a_1 + \lambda a_2) = l(a_1) + \lambda l(a_2)$ over all the vectors of the form $\mathbf{x}_1 \wedge \cdots \wedge \mathbf{x}_p$. Since $\Lambda^p \boldsymbol{E}$ is generated by the vectors of this form (see proposition 4.1), the equality is valid for all the vectors of $\Lambda^p \boldsymbol{E}$. Hence, the linearity of map $l(\cdot)$. Regarding the surjectivity, it suffices to observe that if $L \in (\Lambda^p \boldsymbol{E})^*$, then the map:

$$\phi : \boldsymbol{E}^p \mapsto \mathbb{R}, \quad \varepsilon(\cdot, \cdots, \cdot) := L(\cdot \wedge \cdots \wedge \cdot) \qquad [4.36]$$

is an alternating p-linear form over $\boldsymbol{E}^p$. Now, due to the very construction, it is immediate that $l(\phi) = L$ for all the vectors of the form $\mathbf{x}_1 \wedge \cdots \wedge \mathbf{x}_p$. Since $\Lambda^p \boldsymbol{E}$ is generated by the vectors of this form (see proposition 4.1), then we have $L(\cdot) = l(\phi)(\cdot)$ over $\Lambda^p \boldsymbol{E}$ and thus the map l is indeed surjective. Regarding the injectivity, let us suppose that $l(a_0) = 0$, in this case we have:

$$\forall \mathbf{x}_1, \cdots, \mathbf{x}_p, \quad l(a_0)(\mathbf{x}_1 \wedge \cdots \wedge \mathbf{x}_p) = 0 = a_0(\mathbf{x}_1, \cdots, \mathbf{x}_p), \qquad [4.37]$$

hence the fact that a_0 is null since it is null for all the vectors of $\boldsymbol{E}^p$.

REMARK 4.2.– The proposed proof does not depend on the dimension of the space $\boldsymbol{E}$. Therefore, the isomorphism exists in any case, even if $\boldsymbol{E}$ is not finite dimensional. ■

4.2.4. *Structure of the dual: finite dimensional case*

THEOREM 4.5.– If $\boldsymbol{E}$ is a linear space with finite dimension, then we have the following isomorphisms:

$$(\Lambda^p \boldsymbol{E})^* \cong \mathcal{A}(\boldsymbol{E}^p, \mathbb{R}) \cong (\Lambda^p \boldsymbol{E}^*). \qquad [4.38]$$

Furthermore, the map is defined by

$$\forall (\phi_1, \cdots, \phi_p) \in \boldsymbol{E}^{*p}, \quad \psi_p^* (\phi_1, \cdots, \phi_p) := l (\phi_1 \wedge^* \cdots \wedge^* \phi_p), \qquad [4.39]$$

(where $l : \mathcal{A}(\boldsymbol{E}^p, \mathbb{R}) \mapsto (\Lambda^p \boldsymbol{E})^*$ is the isomorphism of theorem 4.4) is a constructor of the space $\Lambda^p \boldsymbol{E}^*$.

PROOF.– In the case of finite dimension, we have $\dim(\boldsymbol{E}) = \dim(\boldsymbol{E}^*) (= n)$ and for a fixed p, the two spaces $\Lambda^p \boldsymbol{E}, \Lambda^p \boldsymbol{E}^*$ have the same dimension $\binom{n}{p}$. Hence the fact that there exists an isomorphism between $\Lambda^p \boldsymbol{E}$ and $\Lambda^p \boldsymbol{E}^*$. Now, we also know that, in the case of finite dimension, the spaces $\Lambda^p \boldsymbol{E}$ and $(\Lambda^p \boldsymbol{E})^*$ have the same dimension and thus they are isomorphic. Hence, the fact that we have an isomorphism between $(\Lambda^p \boldsymbol{E})^*$ and $(\Lambda^p \boldsymbol{E}^*)$. Furthermore, we have already seen that the space $\mathcal{A}(\boldsymbol{E}^p, \mathbb{R})$ was isomorphic (whichever the dimension) to the space $(\Lambda^p \boldsymbol{E})^*$. Finally, let $\chi_p : \boldsymbol{E}^{*p} \mapsto \Lambda^p \boldsymbol{E}^*$ be a constructor of $\Lambda^p \boldsymbol{E}^*$. There exists a unique linear map κ over $\Lambda^p \boldsymbol{E}^*$ such that:

$$l\left(\psi_p^* (\phi_1, \cdots, \phi_p)\right) = \kappa (\chi_p (\phi_1, \cdots, \phi_p)). \qquad [4.40]$$

Meanwhile, by considering the generated linear spaces, we then have (in virtue of the linearity of κ):

$$\begin{aligned} \left\langle l\left(\psi_p^* (\phi_1, \cdots, \phi_p)\right)\right\rangle &= \langle \kappa (\chi_p (\phi_1, \cdots, \phi_p))\rangle \\ &= \kappa (\langle \chi_p (\phi_1, \cdots, \phi_p)\rangle), \quad (\phi, \cdots, \phi_p) \in \boldsymbol{E}^{*p} \end{aligned} \qquad [4.41]$$

according to proposition 4.1, since χ_p is a constructor of $\Lambda^p \boldsymbol{E}^*$, we know that $\langle \chi_p (\phi_1, \cdots, \phi_p)\rangle = \Lambda^p \boldsymbol{E}^*$. However, we have also seen that we had:

$$\left\langle l\left(\psi_p^* (\phi_1, \cdots, \phi_p)\right)\right\rangle = \Lambda^p \boldsymbol{E}^*. \qquad [4.42]$$

We easily deduce thereof that $\kappa(\Lambda^p \boldsymbol{E}^*) = \Lambda^p \boldsymbol{E}^*$. The map κ is a surjection over $\Lambda^p \boldsymbol{E}^*$ which is finite dimensional. As a result, κ is an isomorphism over $\Lambda^p \boldsymbol{E}^*$. Therefore, $l\left(\psi_p^* (\phi_1, \cdots, \phi_p)\right)$ is also a constructor of $\Lambda^p \boldsymbol{E}^*$.

5

Determinants

At this point, we are capable of constructing bases of $\Lambda^p \boldsymbol{E}$ from bases of $\boldsymbol{E}$. A natural question that we now pose is the following:

QUESTION 5.1.– Let $\boldsymbol{E}$ be a linear space with a finite dimension n. Suppose that we know some vectors $\mathbf{x}_1, \cdots, \mathbf{x}_p$ with their components with respect to the basis $\mathbf{e}_1, \cdots, \mathbf{e}_n$; is it then possible to deduce the components of the vector $\mathbf{x}_1 \wedge \cdots \wedge \mathbf{x}_p$ in a basis $\mathbf{e}_{i_1} \wedge \cdots \wedge \mathbf{e}_{i_p}$?

The answer to this question is, of course, positive. The related tool that we shall introduce is the determinant. Obviously, the properties and the usage of the determinant lie far outside the scope of the calculation of components in the bases of an exterior algebra. However, at the same time, these characteristics help us to realize that a study of exterior algebra makes it possible, in passing, to reintroduce a discourse about the determinants in a natural way.

5.1. Determinant of vector families

5.1.1. *Definition and calculation*

DEFINITION 5.1.– Let $\boldsymbol{E}$ be a linear space of dimension n with a basis $\mathbf{e}_1, \cdots, \mathbf{e}_n$. For all family $\mathbf{x}_1, \cdots, \mathbf{x}_n$ of vectors of $\boldsymbol{E}$ we call the determinant of the family $\mathbf{x}_1, \cdots, \mathbf{x}_n$ in the basis $\mathbf{e}_1, \cdots, \mathbf{e}_n$ the unique real ξ such that:

$$\mathbf{x}_1 \wedge \cdots \wedge \mathbf{x}_n = \xi \mathbf{e}_1 \wedge \cdots \wedge \mathbf{e}_n. \quad [5.1]$$

REMARK 5.1.– This definition requires some commentary:

1) First, the definition is guaranteed, since if $\boldsymbol{E}$ has a dimension n, the space $\Lambda^n \boldsymbol{E}$ has a dimension $\binom{n}{n} = 1$ (see corollary 4.2) with a basis $\mathbf{e}_1 \wedge \cdots \wedge \mathbf{e}_n$ as soon as $\mathbf{e}_1, \cdots, \mathbf{e}_n$ is a basis of $\boldsymbol{E}$. In such a case, it is evident that the determinant of the family actually corresponds to the component of the vector $\mathbf{x}_1 \wedge \cdots \wedge \mathbf{x}_n$ in this basis: we have already a first answer to the introduction's question.

2) *A priori*, in the definition, the designation of the determinant should be subordinated to the symbol "$\wedge$": in fact, what happens if we change the way of "constructing" the exterior product symbol? We shall see that the determinant is actually independent of the chosen symbol of exterior product.

3) Once the basis is chosen, it is clear that the real ξ enables the definition of a function over the set of (ordered) families with n elements within the space $\boldsymbol{E}$ (of dimension n). In principle, we should use the notations:

$$\det_{\mathbf{e}_1, \cdots, \mathbf{e}_n} (\mathbf{x}_1, \cdots, \mathbf{x}_n), \det (\mathbf{x}_1, \cdots, \mathbf{x}_n), \qquad [5.2]$$

according to the desired degree of clarification of the notation. ■

The determinant of a family of vectors can be calculated independently of the exterior product symbol according to the Leibniz formula:

THEOREM 5.1.– [Formula of the Leibniz determinant] Let $\mathbf{e}_1, \cdots, \mathbf{e}_n$ be a basis of the space $\boldsymbol{E}$ and $\mathbf{x}_1, \cdots, \mathbf{x}_n$ a family of n vectors of $\boldsymbol{E}$ with their components written with respect to the basis of $\mathbf{e}_j, j \in [1, n]$ as:

$$\forall i \in [1, n], \ \mathbf{x}_i = \sum_{j=1}^{j=n} x_{j,i} \mathbf{e}_j, \qquad [5.3]$$

in this case, we have:

$$\det_{\mathbf{e}_1, \cdots \mathbf{e}_n} (\mathbf{x}_1, \cdots, \mathbf{x}_n) = \sum_{\sigma \in S_n} \varepsilon(\sigma) x_{\sigma(1),1} \cdots x_{\sigma(n),n}, \qquad [5.4]$$

where we recall that $\varepsilon(\sigma)$ is the signature of the permutation $\sigma \in S_n$.

PROOF.– For this purpose, we employ the multi-linearity of the symbol $\wedge$: we thus have:

$$\mathbf{x}_1 \wedge \cdots \wedge \mathbf{x}_n = \sum_{i_1=1}^{i_1=n} \cdots \sum_{i_n=1}^{i_n=n} x_{i_1,1} \cdots x_{i_n,n} \mathbf{e}_{i_1} \wedge \cdots \wedge \mathbf{e}_{i_n}, \qquad [5.5]$$

since the exterior symbol defines skew-symmetric multi-linear operators, it is clear that we should keep only the terms of summations with the indexes $i_1, \cdots, i_n$ having two by two distinct values. Now, as there are exactly n indexes varying from 1 to n, this is possible only if there exists a permutation $\sigma \in S_n$ such that:

$$\forall k \in [1,n], \ i_k = \sigma(k). \tag{5.6}$$

Therefore, it is possible to write that:

$$\mathbf{x}_1 \wedge \cdots \wedge \mathbf{x}_n = \sum_{\sigma \in S_n} x_{\sigma(1),1} \cdots x_{\sigma(n),n} \mathbf{e}_{\sigma(1)} \wedge \cdots \wedge \mathbf{e}_{\sigma(n)}, \tag{5.7}$$

since the exterior product symbols define an alternating multi-linear operator, we know that:

$$\mathbf{e}_{\sigma(1)} \wedge \cdots \wedge \mathbf{e}_{\sigma(n)} = \varepsilon(\sigma) \mathbf{e}_1 \wedge \cdots \wedge \mathbf{e}_n, \tag{5.8}$$

and we deduce thereof that indeed:

$$\mathbf{x}_1 \wedge \cdots \wedge \mathbf{x}_n = \left[\sum_{\sigma \in S_n} \varepsilon(\sigma) x_{\sigma(1),1} \cdots x_{\sigma(n),n} \right] \mathbf{e}_1 \wedge \cdots \wedge \mathbf{e}_n, \tag{5.9}$$

hence our result. In particular, we can see that the expression of the determinant is independent of the chosen exterior product symbol.

5.1.2. *Components of the exterior product*

In order to be able to work with the components of exterior products of vectors, we now introduce the following definition:

DEFINITION 5.2.– Let $\mathbf{y}$ be a vector of $\boldsymbol{E}$. Let $\mathbf{i} := (i_1, \cdots, i_p) \in \mathcal{I}_o$ be a strictly ordered multi-index of $[1,n]^p$. Let $\mathbf{y}$ be a vector of $\boldsymbol{E}$ decomposed over the basis $\mathbf{e}_1, \cdots, \mathbf{e}_n$ according to $\mathbf{y} = \sum_{i=1}^{i=n} y_i \mathbf{e}_i$. We define $\tilde{\mathbf{y}}(\mathbf{i})$ as the reduced vector of $\mathbf{y}$ by $\mathbf{i}$ in the basis $\mathbf{e}_1, \cdots, \mathbf{e}_n$ according to the expression:

$$\tilde{\mathbf{y}}(\mathbf{i}) := \sum_{l=1}^{l=p} y_{i_l} \mathbf{e}_{i_l}. \tag{5.10}$$

The following theorem allows us to find the components of exterior products of vectors in a given basis $\mathbf{e}_1, \cdots, \mathbf{e}_n$ of $\boldsymbol{E}$.

THEOREM 5.2.– Let $\mathbf{x}_1, \cdots, \mathbf{x}_p$ be a family of p vectors of $\boldsymbol{E}$ and $\mathbf{e}_1, \cdots, \mathbf{e}_n$ a basis of the latter. In this case, we have:

$$\mathbf{x}_1 \wedge \cdots \wedge \mathbf{x}_p = \sum_{\mathbf{i} \in \mathcal{I}_o} \det_{\mathbf{e}_{i_1}, \cdots, \mathbf{e}_{i_p}} \left(\tilde{\mathbf{x}}_1 (\mathbf{i}), \cdots, \tilde{\mathbf{x}}_p (\mathbf{i}) \right) \mathbf{e}_{i_1} \wedge \cdots \wedge \mathbf{e}_{i_p} \qquad [5.11]$$

in lexicon, we call the minor determinant of the family $\mathbf{x}_1, \cdots, \mathbf{x}_p$ for the strictly ordered multi-index $\mathbf{i} \in \mathcal{I}_o$ the quantity:

$$\det_{\mathbf{e}_{i_1}, \cdots, \mathbf{e}_{i_p}} \left(\tilde{\mathbf{x}}_1 (\mathbf{i}), \cdots, \tilde{\mathbf{x}}_p (\mathbf{i}) \right). \qquad [5.12]$$

Before providing the demonstration of this theorem, note the following point:

REMARK 5.2.– The sequence of minor determinants of a family $\mathbf{x}_1, \cdots, \mathbf{x}_p$ for a list of strictly ordered multi-indexes yields exactly the components of the p-vector $\mathbf{x}_1 \wedge \cdots \wedge \mathbf{x}_p$ in the basis of $\mathbf{e}_{i_1} \wedge \cdots \mathbf{e}_{i_p}, \mathbf{i} \in \mathcal{I}_o$, as announced previously. ■

PROOF.– We develop the exterior product:

$$\mathbf{x}_1 \wedge \cdots \wedge \mathbf{x}_p = \sum_{i_1, \cdots, i_p} x_{i_1,1} \cdots x_{i_p,p} \mathbf{e}_{i_1} \wedge \cdots \wedge \mathbf{e}_{i_p} \qquad [5.13]$$

now, it is necessary to rearrange the vectors $\mathbf{e}_{i_1} \wedge \cdots \wedge \mathbf{e}_{i_p}$ in order to keep only the non-null vectors and to group together those that are equal or opposite. This is achieved, of course, by means of the strictly ordered multi-indexes. Consider thus $\mathbf{j} := (j_1, \cdots, j_p)$ a strictly ordered multi-index. Among the multi-indexes (non-necessarily ordered) $\mathbf{i} := i_1, \cdots, i_p$ of the previous summation, there are exactly $p!$ indexes with exactly the same values as $\mathbf{j}$ and thus it can be factorized with respect to the element $\mathbf{j}$. Therefore, we easily obtain that:

$$\mathbf{x}_1 \wedge \cdots \wedge \mathbf{x}_p = \sum_{\mathbf{j} \in \mathcal{I}_o} \sum_{\sigma \in S_p} x_{j_\sigma(1),1} \cdots x_{j_\sigma(p),p} \mathbf{e}_{j_\sigma(1)} \wedge \cdots \wedge \mathbf{e}_{j_\sigma(p)}. \qquad [5.14]$$

By introducing the signatures, it becomes clear that we have:

$$\mathbf{x}_1 \wedge \cdots \wedge \mathbf{x}_p = \sum_{\mathbf{j} \in \mathcal{I}_o} \left(\sum_{\sigma \in S_p} \varepsilon(\sigma) x_{j_\sigma(1),1} \cdots x_{j_\sigma(p),p} \right) \mathbf{e}_{j_1} \wedge \cdots \wedge \mathbf{e}_{j_p}. \qquad [5.15]$$

Next, by employing the expression of a determinant over a family of vectors, as well as the notation from definition 5.2, we obtain exactly the desired theorem.

5.1.3. *Properties of determinants*

PROPOSITION 5.1.– The following properties are true:

1) the map $\det(\cdot,\cdots,\cdot)$ is an alternating n-linear form over the space $\boldsymbol{E}^n$;

2) we have $\det(\mathbf{x}_1,\cdots,\mathbf{x}_n) = 0$ if and only if the family $\mathbf{x}_1,\cdots,\mathbf{x}_n$ is linearly dependent;

3) if $\mathbf{e}_1,\cdots,\mathbf{e}_n$ and $\mathbf{f}_1,\cdots,\mathbf{f}_n$ are two bases of $\boldsymbol{E}$, then we have:

$$\det_{\mathbf{f}_1,\cdots,\mathbf{f}_n}(\mathbf{x}_1,\cdots,\mathbf{x}_n) = \det_{\mathbf{f}_1,\cdots,\mathbf{f}_n}(\mathbf{e}_1,\cdots,\mathbf{e}_n)\det_{\mathbf{e}_1,\cdots,\mathbf{e}_n}(\mathbf{x}_1,\cdots,\mathbf{x}_n). \quad [5.16]$$

PROOF.– Let us demonstrate the properties:

1) It suffices to employ the explicit expression of determinant of the Leibnitz formula [5.1]: let $\alpha \in S_n$ be a permutation over $\{1,\cdots,n\}$. Then, we have:

$$\det\left(\mathbf{x}_{\alpha(1)},\cdots,\mathbf{x}_{\alpha(n)}\right) = \sum_{\sigma\in S_n}\varepsilon(\sigma)\,x_{\sigma\circ\alpha(1),1}\cdots x_{\sigma\circ\alpha(n),n}. \quad [5.17]$$

By performing the "change of variable" $\sigma' \leftrightarrow \sigma\circ\alpha$ within the summation, we obtain:

$$\det\left(\mathbf{x}_{\alpha(1)},\cdots,\mathbf{x}_{\alpha(n)}\right) = \sum_{\sigma'\in S_n}\varepsilon\left(\sigma'\circ\alpha^{-1}\right)x_{\sigma'(1),1}\cdots x_{\sigma'(n),n}, \quad [5.18]$$

we then exploit the property $\varepsilon\left(\sigma'\circ\alpha^{-1}\right) = \varepsilon(\sigma')\,\varepsilon(\alpha)$ to obtain:

$$\det\left(\mathbf{x}_{\alpha(1)},\cdots,\mathbf{x}_{\alpha(n)}\right) =$$
$$\sum_{\sigma'\in S_n}\varepsilon(\sigma')\,\varepsilon(\alpha)\,x_{\sigma'(1),1}\cdots x_{\sigma'(n),n} = \varepsilon(\alpha)\det(\mathbf{x}_1,\cdots,\mathbf{x}_n), \quad [5.19]$$

and we have indeed the fact that the function is alternating. The n-linear aspect is immediate, because the product of n factors is itself an n-linear map.

2) Evident, since for any basis $\mathbf{e}_1,\cdots,\mathbf{e}_n$ we have the equality:

$$\mathbf{x}_1\wedge\cdots\wedge\mathbf{x}_n = \det_{\mathbf{e}_1,\cdots,\mathbf{e}_n}(\mathbf{x}_1,\cdots,\mathbf{x}_n)\,\mathbf{e}_1\wedge\cdots\wedge\mathbf{e}_n. \quad [5.20]$$

Now, the family $\mathbf{x}_1,\cdots,\mathbf{x}_n$ is linearly dependent if and only if $\mathbf{x}_1\wedge\cdots\wedge\mathbf{x}_n$ is null (see theorem 3.2) and thus if and only if its determinant is null.

3) It suffices to develop:

$$\mathbf{x}_1, \cdots, \mathbf{x}_n = \det_{\mathbf{f}_1, \cdots, \mathbf{f}_n}(\mathbf{x}_1, \cdots, \mathbf{x}_n)\, \mathbf{f}_1 \wedge \cdots \wedge \mathbf{f}_n, \qquad [5.21]$$

then, we write simply:

$$\mathbf{f}_1 \wedge \cdots \wedge \mathbf{f}_n = \det_{\mathbf{e}_1, \cdots, \mathbf{e}_n}(\mathbf{f}_1, \cdots, \mathbf{f}_n)\, \mathbf{e}_1 \wedge \cdots \wedge \mathbf{e}_n, \qquad [5.22]$$

then, knowing that:

$$\mathbf{x}_1, \wedge \cdots \wedge \mathbf{x}_n = \det_{\mathbf{e}_1, \cdots, \mathbf{e}_n}(\mathbf{x}_1, \cdots, \mathbf{x}_n)\, \mathbf{e}_1 \wedge \cdots \wedge \mathbf{e}_n, \qquad [5.23]$$

we immediately conclude the proposition.

5.2. Practical calculation of determinants

Within this section, we shall examine different methods of calculating determinants.

5.2.1. *Transposed determinant*

DEFINITION 5.3.– Let $\mathbf{x}_i, i \in [1, n]$ be a family of n vectors of a linear space $\boldsymbol{E}$ of dimension n equipped with the basis $\mathbf{e}_1, \cdots, \mathbf{e}_n$. Suppose that we have written:

$$\forall i \in [1, n], \ \mathbf{x}_i = \sum_{j=1}^{j=n} x_{j,i}\mathbf{e}_j. \qquad [5.24]$$

The transposed family of $\mathbf{x}_i, i \in [1, n]$ is by definition the family $\mathbf{x}_j^*, j \in [1, n]$ such that:

$$\forall j \in [1, n], \mathbf{x}_j^* = \sum_{i=1}^{i=n} x_{j,i}\mathbf{e}_i. \qquad [5.25]$$

Note that the transposition (of family) is an involution: a family is equal to the transposed of its transposed family. We can thus talk about families that are one another's transposed. We can then enunciate the following proposition:

PROPOSITION 5.2.– Let $\mathbf{x}_i, \mathbf{x}_i^*, i \in [1,n]$ be two transposed families (one another's) within the linear space $\boldsymbol{E}$ of dimension n equipped with a basis $\mathbf{e}_1, \cdots, \mathbf{e}_n$. Then, we have:

$$\det_{\mathbf{e}_1,\cdots,\mathbf{e}_n}\left(\mathbf{x}_1^*,\cdots,\mathbf{x}_n^*\right) = \det_{\mathbf{e}_1,\cdots,\mathbf{e}_n}\left(\mathbf{x}_1,\cdots,\mathbf{x}_n\right). \qquad [5.26]$$

PROOF.– It consists simply of a change of variables within the formula of determinants. In fact, we know that, according to the Leibnitz formula (see theorem 5.1), we have:

$$\det_{\mathbf{e}_1,\cdots,\mathbf{e}_n}\left(\mathbf{x}_1,\cdots,\mathbf{x}_n\right) = \sum_{\sigma\in S_n} \varepsilon(\sigma)\, x_{\sigma(1),1}\cdots x_{\sigma(n),n}, \qquad [5.27]$$

which can be written in the form:

$$x_{\sigma(1),1}\cdots x_{\sigma(n),n} = x_{\sigma(1),\sigma^{-1}\sigma(1)}\cdots x_{\sigma(n),\sigma^{-1}\sigma(n)} \qquad [5.28]$$

via a permutation in the multiplication we have, for all permutation σ, the equality:

$$x_{\sigma(1),\sigma^{-1}\sigma(1)}\cdots x_{\sigma(n),\sigma^{-1}\sigma(n)} = x_{1,\sigma^{-1}(1)}\cdots x_{n,\sigma^{-1}(n)}. \qquad [5.29]$$

By employing the fact that the transformation $\sigma \mapsto \sigma^{-1}$ is bijective over the set of permutations, by performing the change of variable $\sigma \rightarrow \sigma^{-1}$ and by using $\varepsilon(\sigma) = \varepsilon\left(\sigma^{-1}\right)$, we naturally obtain:

$$\sum_{\sigma\in S_n} \varepsilon(\sigma)\, x_{\sigma(1),1}\cdots x_{\sigma(n),n} = \sum_{\sigma\in S_n} \varepsilon(\sigma)\, x_{1,\sigma(1)}\cdots x_{n,\sigma(n)}, \qquad [5.30]$$

and the right-hand side term can be easily rewritten as:

$$\sum_{\sigma\in S_n} \varepsilon(\sigma)\, x_{1,\sigma(1)}\cdots x_{n,\sigma(n)} = \det_{\mathbf{e}_1,\cdots,\mathbf{e}_n}\left(\mathbf{x}_1^*,\cdots,\mathbf{x}_n^*\right). \qquad [5.31]$$

5.2.2. *Matrix illustration of determinant calculation*

Let $\mathbf{x}_1, \cdots, \mathbf{x}_n$ be a family of n vectors of $\boldsymbol{E}$, equipped with a basis $\mathbf{e}_1, \cdots, \mathbf{e}_n$. We set that:

$$\forall i \in [1,n], \quad \mathbf{x}_i = \sum_{j=1}^{j=n} x_{j,i}\mathbf{e}_j. \qquad [5.32]$$

Next, we present the components of vectors $\mathbf{x}_i$ according to their components with respect to the basis of $\mathbf{e}_j$ by means of a matrix notation:

$$\begin{array}{ccccc} \mathbf{x}_1 & \mathbf{x}_2 & \cdots & \mathbf{x}_n & \\ x_{11} & x_{12} & \cdots & x_{1n} & \mathbf{e}_1 \\ x_{21} & x_{22} & \cdots & x_{2n} & \mathbf{e}_2 \\ \vdots & \vdots & \ddots & \vdots & \vdots \\ x_{n1} & x_{n2} & \cdots & x_{nn} & \mathbf{e}_n \end{array}, \qquad [5.33]$$

and we set:

$$\det_{\mathbf{e}_1,\cdots,\mathbf{e}_n}(\mathbf{x}_1,\cdots,\mathbf{x}_n) := \begin{vmatrix} x_{11} & x_{12} & \cdots & x_{1n} \\ x_{21} & x_{22} & \cdots & x_{2n} \\ \vdots & \vdots & \ddots & \vdots \\ x_{n1} & x_{n2} & \cdots & x_{nn} \end{vmatrix}. \qquad [5.34]$$

We see then, by means of associating in a bijective way the square matrices of degree n with vectors defined by their components in a base, that it is possible to associate the notion of determinant with any square matrix. The transposed determinant can be represented by inverting the rows and columns of the matrix.

5.2.3. *Laplace expansion along a row or a column*

The Laplace rule is also called the expansion of the determinant along a column (or a row, if we move from a family to its transposed family). In order to write in a legible way, the following notation will be employed:

DEFINITION 5.4.– If $\mathbf{x}_1,\cdots,\mathbf{x}_n$ is a family of n vectors of $\boldsymbol{E}$ and if $\mathbf{x}_{i_1},\cdots,\mathbf{x}_{i_p}$ is a family of p vectors with $\mathbf{i} \in \mathcal{I}_o(p)$ a strictly ordered p-index, we shall denote by:

$$\underbrace{\mathbf{x}_1 \wedge \cdots \wedge \mathbf{x}_n}_{\textit{omitted } \mathbf{x}_{i_1},\cdots,\mathbf{x}_{i_p}} := \left(\mathbf{x}_1 \wedge \cdots \wedge \mathbf{x}_n \backslash \left[\mathbf{x}_{i_1},\cdots,\mathbf{x}_{i_p}\right]\right). \qquad [5.35]$$

In order to introduce the Laplace formula, let us choose $l \in [1,n]$, develop $\mathbf{x}_l$ with respect to the basis of $\mathbf{e}_k$, $k \in [1,n]$ and then calculate according to the multi-linearity of the exterior product symbol:

$$\mathbf{x}_1 \wedge \cdots \wedge \mathbf{x}_n = \sum_{k=1}^{k=n} x_{k,l} \left(\mathbf{x}_1 \wedge \cdots \wedge \underbrace{\mathbf{e}_k}_{\textit{position } l} \wedge \cdots \wedge \mathbf{x}_n \right)$$

$$= \sum_{k=1}^{k=n} x_{k,l} (-1)^{n-l} (\mathbf{x}_1 \wedge \cdots \wedge \mathbf{x}_n \backslash [\mathbf{x}_l]) \wedge \mathbf{e}_k. \qquad [5.36]$$

For a k fixed in $[1, n]$, we consider now the strictly ordered n-index:

$$\boldsymbol{\kappa} = (\kappa_1, \cdots \kappa_{n-1}),$$
$$\textit{such that } \kappa_1 = 1, \cdots \kappa_{k-1} = k-1, \kappa_k = k+1, \cdots, \kappa_{n-1} = n. \qquad [5.37]$$

This is the strictly ordered complementary of k in the set $[1, n]$. Then, for each $k \in [1, n]$, we have:

$$(\mathbf{x}_1 \wedge \cdots \wedge \mathbf{x}_n \backslash [\mathbf{x}_l]) \wedge \mathbf{e}_k$$
$$= \det [(\tilde{\mathbf{x}}_1 (\boldsymbol{\kappa}) \wedge \cdots \wedge \tilde{\mathbf{x}}_n (\boldsymbol{\kappa}) \backslash [\tilde{\mathbf{x}}_l (\boldsymbol{\kappa})])] \mathbf{e}_{\kappa_1} \wedge \cdots \wedge \mathbf{e}_{\kappa_{n-1}} \wedge \mathbf{e}_k. \qquad [5.38]$$

Clearly, we have:

$$\mathbf{e}_{\kappa_1} \wedge \cdots \wedge \mathbf{e}_{\kappa_{n-1}} \wedge \mathbf{e}_k = (-1)^{n-k} \mathbf{e}_1 \wedge \cdots \wedge \mathbf{e}_n. \qquad [5.39]$$

Hence, we finally deduce:

$$\mathbf{x}_1 \wedge \cdots \wedge \mathbf{x}_n = \sum_{k=1}^{k=n} (-1)^{n-l} (-1)^{n-k} x_{k,l}$$
$$\det [(\tilde{\mathbf{x}}_1 (\boldsymbol{\kappa}) \wedge \cdots \wedge \tilde{\mathbf{x}}_n (\boldsymbol{\kappa}) \backslash [\tilde{\mathbf{x}}_l (\boldsymbol{\kappa})])] \mathbf{e}_1 \wedge \cdots \wedge \mathbf{e}_n. \qquad [5.40]$$

By noting that $(-1)^{n-l} (-1)^{n-k} = (-1)^{k+l}$, we then have:

$$\det (\mathbf{x}_1, \cdots, \mathbf{x}_n) = \sum_{k=1}^{k=n} (-1)^{k+l} x_{k,l} \det [(\tilde{\mathbf{x}}_1 (\boldsymbol{\kappa}) \wedge \cdots \wedge \tilde{\mathbf{x}}_n (\boldsymbol{\kappa}) \backslash [\tilde{\mathbf{x}}_l (\boldsymbol{\kappa})])], \qquad [5.41]$$

which is the Laplace formula of the determinant with respect to the column l:

THEOREM 5.3 (Laplace formula of expansion with respect to a row).– Let $\mathbf{x}_1, \cdots, \mathbf{x}_n$ be a family of n vectors of $\boldsymbol{E}$. For all $k \in [1, n]$, we set the strictly ordered $n-1$-index dependent on k:

$$\boldsymbol{\kappa} = (\kappa_1, \cdots \kappa_{n-1}) \in \mathcal{I}_o,$$

$$\textit{such that } \kappa_1 = 1, \cdots \kappa_{k-1} = k-1, \kappa_k = k+1, \cdots, \kappa_{n-1} = n. \quad [5.42]$$

Then, for all $l \in [1, n]$, it is possible to expand the determinant according to the formula:

$$\det(\mathbf{x}_1, \cdots, \mathbf{x}_n)$$
$$= \sum_{k=1}^{k=n} (-1)^{k+l} x_{k,l} \det\left[(\tilde{\mathbf{x}}_1(\boldsymbol{\kappa}) \wedge \cdots \wedge \tilde{\mathbf{x}}_n(\boldsymbol{\kappa}) \setminus [\tilde{\mathbf{x}}_l(\boldsymbol{\kappa})])\right]. \quad [5.43]$$

5.2.4. *Expansion by blocks*

THEOREM 5.4 (Calculation of determinants by blocks).– Consider a family $\mathbf{x}_1, \cdots, \mathbf{x}_n$ of n vectors of $\boldsymbol{E}$ equipped with a basis $\mathbf{e}_1, \cdots, \mathbf{e}_n$. We suppose that there exists an integer $p \in [1, n-1]$ such that:

$$\forall k \le p, \ \mathbf{x}_k = \sum_{j=1}^{j=p} x_{j,k} \mathbf{e}_j. \quad [5.44]$$

We then consider the strictly ordered $n-p$-index $\mathbf{i} := (p+1, \cdots, n) \in \mathcal{I}_o(n-p)$ and the family of reduced vectors $\tilde{\mathbf{x}}_{p+1}(\mathbf{i}), \cdots, \tilde{\mathbf{x}}_n(\mathbf{i})$, we then have the equality:

$$\det_{\mathbf{e}_1, \cdots, \mathbf{e}_n}(\mathbf{x}_1, \cdots, \mathbf{x}_n)$$
$$= \det_{\mathbf{e}_1, \cdots, \mathbf{e}_p}(\mathbf{x}_1, \cdots, \mathbf{x}_p) \det_{\mathbf{e}_{p+1}, \cdots, \mathbf{e}_n}(\tilde{\mathbf{x}}_{p+1}(\mathbf{i}), \cdots, \tilde{\mathbf{x}}_n(\mathbf{i})). \quad [5.45]$$

PROOF.– The proof is quite immediate within the formalism of exterior algebras. Indeed, due to the particular form of vectors $\mathbf{x}_1, \cdots, \mathbf{x}_p$, we calculate at first:

$$\mathbf{x}_1 \wedge \cdots \wedge \mathbf{x}_p = \det_{\mathbf{e}_1, \cdots, \mathbf{e}_p}(\mathbf{x}_1, \cdots, \mathbf{x}_p) \mathbf{e}_1 \wedge \cdots \wedge \mathbf{e}_p. \quad [5.46]$$

Let us now develop the exterior product of the other vectors:

$$\mathbf{x}_{p+1} \wedge \cdots \wedge \mathbf{x}_n = (\mathbf{x}_{p+1} - \tilde{\mathbf{x}}_{p+1}(\mathbf{i}) + \tilde{\mathbf{x}}_{p+1}(\mathbf{i})) \wedge \cdots \wedge (\mathbf{x}_n - \tilde{\mathbf{x}}_n(\mathbf{i}) + \tilde{\mathbf{x}}_n(\mathbf{i})), \quad [5.47]$$

it is clear, in virtue of the very construction, that each vector of the form $\mathbf{x}_j - \tilde{\mathbf{x}}_j(\mathbf{i})$, $j \geq p+1$ admits only the components over the vectors $\mathbf{e}_1, \cdots, \mathbf{e}_p$. Therefore, by developing the term above, thanks to the multi-linearity of symbols $\wedge$, there will be only one symbol without terms over $\mathbf{e}_1, \cdots, \mathbf{e}_p$, that is:

$$\tilde{\mathbf{x}}_{p+1}(\mathbf{i}) \wedge \cdots \wedge \tilde{\mathbf{x}}_n(\mathbf{i}) = \det_{\mathbf{e}_{p+1}, \cdots, \mathbf{e}_n} (\tilde{\mathbf{x}}_{p+1}(\mathbf{i}), \cdots, \tilde{\mathbf{x}}_n(\mathbf{i})) \, \mathbf{e}_{p+1} \wedge \cdots \wedge \mathbf{e}_n. \quad [5.48]$$

Therefore, we immediately obtain:

$$\mathbf{x}_1 \wedge \cdots \wedge \mathbf{x}_p \wedge (\mathbf{x}_{p+1} \wedge \cdots \wedge \mathbf{x}_n)$$
$$= \det_{\mathbf{e}_1, \cdots, \mathbf{e}_p} (\mathbf{x}_1, \cdots, \mathbf{x}_p) \, \mathbf{e}_1 \wedge \cdots \mathbf{e}_p \wedge (\tilde{\mathbf{x}}_{p+1}(\mathbf{i}) \wedge \cdots \wedge \tilde{\mathbf{x}}_n(\mathbf{i})). \quad [5.49]$$

Hence the announced result, by replacing $\tilde{\mathbf{x}}_{p+1}(\mathbf{i}) \wedge \cdots \wedge \tilde{\mathbf{x}}_n(\mathbf{i})$ with formula [5.48].

5.2.5. *Expansion by exploiting the associativity*

The calculation above is an interesting case since it splits the determinant into sub-calculations. First, an exterior product of several vectors is calculated, and then we calculate the exterior product of the remaining vectors, and by recomposing the total exterior product, we obtain the formula for the desired determinant. Therefore, the associativity of the exterior product can allow us to "factorize" the calculation of determinant. Besides some computational advantages that this method surely provides, it also has clear advantages in practical examples for the fans of mental calculation.

Case of dimension three. Consider the family of three vectors $\mathbf{x}_1, \mathbf{x}_2, \mathbf{x}_3$, given by:

$$\mathbf{x}_1 = 1\mathbf{e}_1 - 2\mathbf{e}_2 + 2\mathbf{e}_3, \quad \mathbf{x}_2 = -\mathbf{e}_1 + 1\mathbf{e}_3, \quad \mathbf{x}_3 = 2\mathbf{e}_1 + 1\mathbf{e}_2 - \mathbf{e}_3, \quad [5.50]$$

that can be visualized with the following matrix:

$$\begin{bmatrix} 1 & -1 & 2 \\ -2 & 0 & 1 \\ 2 & 1 & -1 \end{bmatrix}. \quad [5.51]$$

In order to calculate the determinant of this family of vectors in the basis $\mathbf{e}_1, \mathbf{e}_2, \mathbf{e}_3$, it is necessary to find the coordinate of $\mathbf{x}_1 \wedge \mathbf{x}_2 \wedge \mathbf{x}_3$ in the basis $\mathbf{e}_1 \wedge \mathbf{e}_2 \wedge \mathbf{e}_3$. Thanks to the associativity of the exterior product, we do not need to calculate the triple product directly. At first, the bivector $\mathbf{x}_1 \wedge \mathbf{x}_2$ is calculated. Thus, we place the two vectors $\mathbf{x}_1$ and $\mathbf{x}_2$ as two columns next to each other. To calculate the component of the vector $\mathbf{x}_1 \wedge \mathbf{x}_2$ with respect to the vector $\mathbf{e}_i \wedge \mathbf{e}_j, i < j$, we follow the "$\gamma$" rule between the components of these vectors along the rows i and j: we calculate $x_{i,1}x_{j,2} - x_{j,1}x_{i,2}$. The intermediate calculation is then presented in the following way:

$$\mathbf{x}_1 \wedge \mathbf{x}_2 = \begin{bmatrix} -2\ \mathbf{e}_1 \wedge \mathbf{e}_2 \\ 3\ \ \mathbf{e}_1 \wedge \mathbf{e}_3 \\ -2\ \mathbf{e}_2 \wedge \mathbf{e}_3 \end{bmatrix} \quad \mathbf{x}_3 = \begin{bmatrix} 2\mathbf{e}_1 \\ 1\mathbf{e}_2 \\ -1\mathbf{e}_3 \end{bmatrix}. \qquad [5.52]$$

In order to complete the calculation, we perform the product of $(\mathbf{x}_1 \wedge \mathbf{x}_2)$ (which is thus already known) with the vector $\mathbf{x}_3$. For an easier presentation of the calculations, we proceed in the following way:

1) the components of the vector $\mathbf{x}_1 \wedge \mathbf{x}_2$ are kept, along the column, in the same order in which they appear;

2) by means of $\mathbf{x}_3$, a new column vector is created by inverting (between the top and the bottom) the order of components: the first component becomes the last and vice versa, the second component becomes the penultimate and vice versa, and so forth;

3) the (ordered) components of $\mathbf{x}_1 \wedge \mathbf{x}_2$ and the (inverted) components of $\mathbf{x}_3$ are placed side by side;

4) a third sign column is created (i.e. +1 or −1 which corresponds in fact to the signature of a permutation) in the following way: the filling is done symmetrically between the "top" and the "bottom", starting with a sign + and then alternating the + and the −.

Visually, this approach yields:

$$\begin{array}{ccc} \mathbf{x}_1 \wedge \mathbf{x}_2 & \textit{inverted}\,(\mathbf{x}_3) & \textit{sign} \\ -2 & -1 & + \\ 3 & 1 & - \\ -2 & 2 & + \end{array} \qquad [5.53]$$

Then, the determinant calculation is performed:

$$\det_{\mathbf{e}_1,\mathbf{e}_2\mathbf{e}_3}(\mathbf{x}_1, \mathbf{x}_2\mathbf{x}_3) = +(-2)(-1) - (3)(1) + (-2)(2) = -5. \qquad [5.54]$$

Case of dimension 4. Consider the family of 4 vectors of $\mathbb{R}^4$:

$$\begin{aligned} \mathbf{x}_1 &= \mathbf{e}_1 - 2\mathbf{e}_2 - \mathbf{e}_3 + 2\mathbf{e}_4, \\ \mathbf{x}_2 &= -\mathbf{e}_1 + 2\mathbf{e}_2 + 2\mathbf{e}_4, \\ \mathbf{x}_3 &= 2\mathbf{e}_1 + \mathbf{e}_3 - 1\mathbf{e}_4, \\ \mathbf{x}_4 &= -2\mathbf{e}_1 + \mathbf{e}_2 - 1\mathbf{e}_3 + 2\mathbf{e}_4. \end{aligned} \quad [5.55]$$

which are represented in the form of a matrix:

$$\begin{bmatrix} 1 & -1 & 2 & -2 \\ -2 & 2 & 0 & 1 \\ -1 & 0 & 1 & -1 \\ 2 & -2 & -1 & 2 \end{bmatrix}. \quad [5.56]$$

The method is then identical to the previous approach:

1) the bivectors $\mathbf{x}_1 \wedge \mathbf{x}_2$ and $\mathbf{x}_3 \wedge \mathbf{x}_4$ are calculated in the basis (lexicographically ordered) $\mathbf{e}_i \wedge \mathbf{e}_j, i < j$;

2) starting from the bivector column $\mathbf{x}_3 \wedge \mathbf{x}_4$, a new bivector column is created by inverting (between the top and the bottom) the order of components: the first component becomes the last and vice versa, the second component becomes the penultimate and vice versa, and so forth;

3) the (ordered) components of $\mathbf{x}_1 \wedge \mathbf{x}_2$ and the (inverted) components of $\mathbf{x}_3 \wedge \mathbf{x}_4$ are placed side by side;

4) a third sign column is created (i.e. +1 or −1 which corresponds in fact to the signature of a permutation) in the following way: the filling is done symmetrically between the "top" and the "bottom", starting with a sign + and then alternating the + and the −.

In this way, the following presentation is obtained:

$$\begin{bmatrix} 0\mathbf{e}_1 \wedge \mathbf{e}_2 & 2\mathbf{e}_1 \wedge \mathbf{e}_2 \\ -1\mathbf{e}_1 \wedge \mathbf{e}_3 & 0\mathbf{e}_1 \wedge \mathbf{e}_3 \\ 0\mathbf{e}_1 \wedge \mathbf{e}_4 & 2\mathbf{e}_1 \wedge \mathbf{e}_4 \\ 2\mathbf{e}_2 \wedge \mathbf{e}3 & -1\mathbf{e}_2 \wedge \mathbf{e}3 \\ 0\mathbf{e}_2 \wedge \mathbf{e}_4 & 1\mathbf{e}_2 \wedge \mathbf{e}_4 \\ 2\mathbf{e}_3 \wedge \mathbf{e}_4 & 1\mathbf{e}_3 \wedge \mathbf{e}_4 \end{bmatrix} \rightarrow \begin{matrix} \mathbf{x}_1 \wedge \mathbf{x}_2 & \textit{inverted}\,(\mathbf{x}_3 \wedge \mathbf{x}_4) & \textit{sign} \\ 0 & 1 & + \\ -1 & 1 & - \\ 0 & -1 & + \\ 2 & 2 & + \\ 0 & 0 & - \\ 2 & 2 & + \end{matrix} \quad [5.57]$$

We then obtain:

$$\det_{\mathbf{e}_1,\cdots,\mathbf{e}_4}(\mathbf{x}_1,\cdots,\mathbf{x}_4) =$$
$$+(0)(1)-(-1)(1)+(0)(-1)+(2)(2)-(0)(0)+(2)(2)=9. \quad [5.58]$$

REMARK 5.3.– 1) We shall discover further in this book where this rule comes from. Basically, it employs calculation properties related to the Hodge conjugation.

2) In dimension four, the proposed method is faster than the plain formula. In fact, the latter requires performing a summation over the permutations, and therefore for each permutation 4 multiplications have to be done, which yields a total of 96 k-flops (i.e. the cost of an elementary operation). That requires: 18 operations for the calculation of each bivector (i.e. 36 operations) plus another 12 operations for the sums of products. Thus, 48 operations are required.

3) Without any doubt, for the case of dimension five, it is possible to propose calculation methods simple enough to be organized in writing, in order to calculate the determinants in a relatively elementary way without employing a computer.

4) The general question of determinant calculation by means of associations of exterior products together with the question of minimal complexity seems to be a logical issue. In particular, in the case of sparse matrices, the exterior products do not appear to be too huge to be calculated, which makes the algebraic approach feasible. ■

5.3. Solution of linear systems

5.3.1. *Cramer formulas*

The exterior product will allow us to solve linear systems in a formal way. For this purpose, consider a linear system written in the form:

$$\forall i \in [1,n], \quad \sum_{j=1}^{j=n} a_{ij}x_j = b_i, \quad [5.59]$$

where the coefficients $a_{ij}, i,j \in [1,n]$ are given and the right-hand side terms $b_i, i \in [1,n]$ are known as well, with $x_j, j \in [1,n]$ unknown. In all that follows, we denote:

$$\mathbf{a}_j = \sum_{i=1}^{i=n} a_{ij}\mathbf{e}_i, \quad \mathbf{b} = \sum_{i=1}^{i=n} b_i\mathbf{e}_i, \quad [5.60]$$

where $\mathbf{e}_1, \cdots, \mathbf{e}_n$ is a basis of $\boldsymbol{E}$. We suppose that the family $\mathbf{a}_j, j \in [1, n]$ is a linearly independent family of $\boldsymbol{E}$. It is possible to rewrite the system of equations in the form:

$$\sum_{j=1}^{j=n} x_j \mathbf{a}_j = \mathbf{b}, \qquad [5.61]$$

we select $k \in [1, n]$ and we multiply both sides of the equality by the vector $\mathbf{a}_1 \wedge \cdots \wedge \mathbf{a}_n / [\mathbf{a}_k]$, thus obtaining formally:

$$\left(\sum_{j=1}^{j=n} x_j \mathbf{a}_j \right) \wedge (\mathbf{a}_1 \wedge \cdots \wedge \mathbf{a}_n / [\mathbf{a}_k]) = \mathbf{b} \wedge \mathbf{a}_1 \wedge \cdots \wedge \mathbf{a}_n / [\mathbf{a}_k]. \qquad [5.62]$$

Within the left-hand side, by distributing the exterior product over the summation and by removing the products in which the term $\mathbf{a}_k$ appears twice, we obtain:

$$x_k \mathbf{a}_k \wedge (\mathbf{a}_1 \wedge \cdots \wedge \mathbf{a}_n / [\mathbf{a}_k]) = \mathbf{b} \wedge (\mathbf{a}_1 \wedge \cdots \wedge \mathbf{a}_n / [\mathbf{a}_k]), \qquad [5.63]$$

which can be rewritten, up to some permutation:

$$x_k \mathbf{a}_1 \wedge \cdots \wedge \mathbf{a}_n = \mathbf{a}_1 \wedge \cdots \wedge \mathbf{a}_{k-1} \wedge \mathbf{b} \wedge \mathbf{a}_{k+1} \wedge \cdots \wedge \mathbf{a}_n. \qquad [5.64]$$

Thus, we can see that it is possible to find the value of x_k in the following way: first, the determinant of the family of $\mathbf{a}_i, i \in [1, n]$ is calculated, and then we calculate the determinant of the family of $\mathbf{a}_i$ where the vector (column) $\mathbf{a}_k$ has been replaced by the vector (column) $\mathbf{b}$. The ratio between the second determinant and the first one (which is non-null, since the family is linearly independent) then yields the value of x_k: this is the ***Cramer*** formula. It is evident, according to formula [5.64], that the value of x_k is the component of the vector $\mathbf{b}$ over the vector $\mathbf{a}_k$ in the basis of $\mathbf{a}_i, i \in [1, n]$.

5.3.2. *Rouché-Fontené theorem*

In this section, we describe the Rouché-Capelli theorem (also known as Rouché-Fontené in France, Kronecker-Capelli in Poland and Russia, and Rouché-Frobenius in Spain and Latin America) by means of the exterior formalism, i.e. by describing the three points below, since the theorem concerns a linear system:

1) a necessary and sufficient condition for the existence of solutions;

2) in the case of existence of solutions, a necessary and sufficient condition for uniqueness;

3) in the case of existence and of non-uniqueness, a complete description of solutions.

Let us start with a necessary and sufficient condition for the existence:

THEOREM 5.5.– Consider a linear system written in vector form as:

$$\sum_{j=1}^{j=m} x_j \mathbf{a}_j = \mathbf{b}, \quad \mathbf{a}_j = \sum_{i=1}^{i=n} a_{ij}\mathbf{e}_i, \quad \mathbf{b} = \sum_{i=1}^{i=n} b_i \mathbf{e}_i, \tag{5.65}$$

where $\mathbf{e}_1, \cdots, \mathbf{e}_n$ is a basis of a space $\boldsymbol{E}$ with a dimension n. We suppose that not all the $\mathbf{a}_i, i \in [1, n]$ are null. Then, the linear system admits (at least) a solution if and only if there exists an ordered multi-index $\mathbf{i} := (i_1, \cdots, i_r) \in \mathcal{I}_o$ such that:

$$\mathbf{a}_{i_1} \wedge \cdots \wedge \mathbf{a}_{i_r} \neq \mathbf{0} \textit{ and } \mathbf{b} \wedge \mathbf{a}_{i_1} \wedge \cdots \wedge \mathbf{a}_{i_r} = \mathbf{0}. \tag{5.66}$$

PROOF.– Suppose that we have found $\mathbf{i} = (i_1, \cdots i_r)$ such that:

$$\mathbf{a}_{i_1} \wedge \cdots \wedge \mathbf{a}_{i_r} \neq \mathbf{0} \tag{5.67}$$

then, this means that the family $\mathbf{a}_{i_1}, \cdots \mathbf{a}_{i_r}$ is a linearly independent family. As a result, we have:

$$\mathbf{b} \wedge \mathbf{a}_{i_1} \wedge \cdots \wedge \mathbf{a}_{i_r} = \mathbf{0} \tag{5.68}$$

if and only if $\mathbf{b}$ belongs to the subspace generated by $\mathbf{a}_{i_1}, \cdots, \mathbf{a}_{i_r}$. By definition, there exist $x_{i_1}, \cdots x_{i_r}$ such that:

$$\sum_{k=1}^{k=r} x_{i_k} \mathbf{a}_{i_k} = \mathbf{b} \tag{5.69}$$

then, if we pose all the $x_j, j \notin \{i_1, \cdots, i_r\}$ null, the following equality is guaranteed:

$$\sum_{j=1}^{j=m} x_j \mathbf{a}_j = \mathbf{b}. \tag{5.70}$$

Vice versa, suppose that there exist some scalars $x_j, j \in [1, m]$ that satisfy equality [5.70]. Then, this means that the vector $\mathbf{b}$ belongs to the space $\boldsymbol{A}$ generated by the $\mathbf{a}_i, i \in \{1, m\}$. It is possible to extract from this generating family a basis of $\boldsymbol{A}$, which we denote by $\mathbf{a}_{i_1}, \cdots \mathbf{a}_{i_r}$. As a result, we have indeed $\mathbf{a}_{i_1} \wedge \cdots \wedge \mathbf{a}_{i_r} \neq \mathbf{0}$ and $\mathbf{b} \wedge \mathbf{a}_{i_1} \wedge \cdots \wedge \mathbf{a}_{i_r} = \mathbf{0}$. The reciprocal condition is therefore demonstrated.

Let us now consider a necessary and sufficient condition for uniqueness:

THEOREM 5.6.– Consider a linear system written in vector form:

$$\sum_{j=1}^{j=m} x_j \mathbf{a}_j = \mathbf{b}. \qquad [5.71]$$

We suppose that this system admits at least one solution (see the characterization according to the previous theorem). In this hypothesis, the solution is unique if and only if the family $\mathbf{a}_1, \cdots, \mathbf{a}_m$ is a linearly independent family.

PROOF.– Suppose that the linear system admits a solution and that the family is linearly independent. Then, since $\mathbf{a}_1, \cdots, \mathbf{a}_m$ is a basis of $\boldsymbol{A} := \langle \mathbf{a}_1, \cdots, \mathbf{a}_m \rangle$, and since, according to formula [5.71] the vector $\mathbf{b}$ is in $\boldsymbol{A}$, this means that $\mathbf{b}$ can be written in a unique way:

$$\mathbf{b} = \sum_{j=1}^{j=m} b_j \mathbf{a}_j, \qquad [5.72]$$

hence we deduce that $\forall j \in [1, m], x_j = b_j$, which guarantees the uniqueness of the coefficients x_j and thus the uniqueness of the solutions of the linear system. Vice versa, in the hypothesis that there exists a solution, suppose that this solution is unique. If the family of $\mathbf{a}_j, j \in [1, m]$ was not linearly independent, we could find a null linear combination where not all coefficients are null:

$$\sum_{j=1}^{j=m} \nu_j \mathbf{a}_j = \mathbf{0} \textit{ and } \exists j \textit{ such that } \nu_j \neq 0. \qquad [5.73]$$

Let therefore $x_j, j \in [1, m]$ be a solution of the system [5.71]. Clearly, the family $y_j := x_j + \nu_j, j \in [1, m]$ is still a solution of the linear system. However, there exists at least one y_j which is not equal to x_j (since there exists at least one ν_j non-null). As a result, we have at least two distinct solutions of the linear system, which is in contradiction with our hypothesis. Therefore, the family of $\mathbf{a}_j, j \in [1, m]$ is necessarily a linearly independent family.

Finally, we conclude with the characterization of the writing of solutions in the case of a linearly dependent family:

THEOREM 5.7.– Consider a linear system written in vector form:

$$\sum_{j=1}^{j=m} x_j \mathbf{a}_j = \mathbf{b}. \quad [5.74]$$

Suppose that this system admits more than one solution (in other words, the family $\mathbf{a}_1, \cdots, \mathbf{a}_m$ is not linearly independent, see the characterization of the previous theorem). In this case, we define by $\mathbf{a}_{i_1}, \cdots, \mathbf{a}_{i_r}$ a basis of a linear space $\boldsymbol{A} = \langle \mathbf{a}_1, \cdots, \mathbf{a}_m \rangle$. The linear system can then be rewritten in the form:

$$\sum_{k=1}^{k=r} x_{i_k} \mathbf{a}_{i_k} = \left(\mathbf{b} - \sum_{j \notin \{i_1, \cdots, i_r\}} x_j \mathbf{a}_j \right). \quad [5.75]$$

Then, for any family $x_j, j \notin \{i_1, \cdots, i_r\}$, the system above admits a unique solution in $x_{i_k}, k \in [1, r]$ and this solution process describes the whole set of solutions of the linear system.

PROOF.– Suppose that the system [5.74] admits some solutions. Then, there exists a family $\mathbf{a}_{j_1}, \cdots, \mathbf{a}_{j_s}$ of vectors such that:

$$\mathbf{a}_{j_1} \wedge \cdots \wedge \mathbf{a}_{j_s} \neq \mathbf{0} \textit{ and } \mathbf{b} \wedge \mathbf{a}_{j_1} \wedge \cdots \wedge \mathbf{a}_{j_s} = \mathbf{0}, \quad [5.76]$$

it is possible to complete the list $\mathbf{a}_{j_1}, \cdots \mathbf{a}_{j_s}$ by some vectors $\mathbf{a}_{j_{s+1}}, \cdots \mathbf{a}_{j_r}$ in order to construct a basis of $\boldsymbol{A}$. In particular, we can write that:

$$\mathbf{a}_{j_1} \wedge \cdots \wedge \mathbf{a}_{j_r} = \alpha \mathbf{a}_{i_1} \wedge \cdots \wedge \mathbf{a}_{i_r}, \ \alpha \neq 0. \quad [5.77]$$

Thus, we have necessarily:

$$\mathbf{b} \wedge \mathbf{a}_{j_1} \wedge \cdots \wedge \mathbf{a}_{j_s} = \mathbf{0} \Rightarrow \mathbf{b} \wedge \mathbf{a}_{j_1} \wedge \cdots \wedge \mathbf{a}_{j_r} = \mathbf{0}$$
$$\Rightarrow \mathbf{b} \wedge \mathbf{a}_{i_1} \wedge \cdots \wedge \mathbf{a}_{i_r} = \mathbf{0}. \quad [5.78]$$

1) Let us arbitrarily fix the reals $x_j, j \notin \{i_1, \cdots, i_r\}$ and consider the linear system [5.75]. Clearly, according to the preliminary calculation above, the right-hand side satisfies:

$$\left(\mathbf{b} - \sum_{j \notin \{i_1, \cdots, i_r\}} x_j \mathbf{a}_j \right) \wedge \mathbf{a}_{i_1} \wedge \cdots \wedge \mathbf{a}_{i_r} = \mathbf{0}, \quad [5.79]$$

since $\mathbf{a}_{i_1} \wedge \cdots \wedge \mathbf{a}_{i_r} \neq \mathbf{0}$, this entails, in virtue of theorem 5.5, that there exists a solution for the unknown $x_{i_1}, \cdots x_{i_r}$. Furthermore, since the family $\mathbf{a}_{i_1}, \cdots, \mathbf{a}_{i_r}$ is linearly independent, this means, in virtue of theorem 5.6, that the solution is unique.

2) Vice versa, suppose we know a solution of the linear system [5.74]. Then, by moving the $j \notin \{i_1, \cdots, i_r\}$ to the right-hand side, we obtain a relation of the form [5.75], and the $x_{i_1}, \cdots, x_{i_r}$ that satisfy such a relation are necessarily unique, due to the above argument.

5.3.3. *Example*

Let us illustrate an example with $n = 4$, $m = 3$ and $r = 2$. Consider the following vectors:

$$\begin{aligned} \mathbf{a}_1 &= 1\mathbf{e}_1 - 1\mathbf{e}_2 + 0\mathbf{e}_3 + 1\mathbf{e}_4 \\ \mathbf{a}_2 &= 1\mathbf{e}_1 + 1\mathbf{e}_2 + 1\mathbf{e}_3 + 0\mathbf{e}_4 \\ \mathbf{a}_3 &= 3\mathbf{e}_1 - 1\mathbf{e}_2 + 1\mathbf{e}_3 + 2\mathbf{e}_4 \\ \mathbf{b} &= -1\mathbf{e}_1 + 3\mathbf{e}_2 + 1\mathbf{e}_3 + t\mathbf{e}_4 \end{aligned} \qquad [5.80]$$

where t is at first an undefined value. On the other hand, it can be easily observed that $\mathbf{a}_3 = 2\mathbf{a}_1 + \mathbf{a}_2$, whereas $\mathbf{a}_1, \mathbf{a}_2$ are linearly independent. In the first phase, we verify the existence of a solution, by proceeding in an incremental way:

$$\mathbf{a}_1 \neq \mathbf{0}, \;\; \mathbf{b} \wedge \mathbf{a}_1 = \left| \begin{array}{rl} 2 & \mathbf{e}_1 \wedge \mathbf{e}_2 \\ 1 & \mathbf{e}_1 \wedge \mathbf{e}_3 \\ t+1 & \mathbf{e}_1 \wedge \mathbf{e}_4 \\ -1 & \mathbf{e}_2 \wedge \mathbf{e}_3 \\ -t-3 & \mathbf{e}_2 \wedge \mathbf{e}_4 \\ -1 & \mathbf{e}_3 \wedge \mathbf{e}_4 \end{array} \right. \qquad [5.81]$$

Evidently, it is never possible to have $\mathbf{b} \wedge \mathbf{a}_1 = \mathbf{0}$. Therefore, we can add the vector $\mathbf{a}_2$ to the exterior products:

$$\mathbf{a}_1 \wedge \mathbf{a}_2 = \left| \begin{array}{rl} 2 & \mathbf{e}_1 \wedge \mathbf{e}_2 \\ 1 & \mathbf{e}_1 \wedge \mathbf{e}_3 \\ -1 & \mathbf{e}_1 \wedge \mathbf{e}_4 \\ -1 & \mathbf{e}_2 \wedge \mathbf{e}_3 \\ -1 & \mathbf{e}_2 \wedge \mathbf{e}_4 \\ -1 & \mathbf{e}_3 \wedge \mathbf{e}_4 \end{array} \right. , \quad \mathbf{b} \wedge \mathbf{a}_1 \wedge \mathbf{a}_2 = \left| \begin{array}{rl} 0 & \mathbf{e}_1 \wedge \mathbf{e}_2 \wedge \mathbf{e}_3 \\ -2t-4 & \mathbf{e}_1 \wedge \mathbf{e}_2 \wedge \mathbf{e}_4 \\ -t-2 & \mathbf{e}_1 \wedge \mathbf{e}_3 \wedge \mathbf{e}_4 \\ t+2 & \mathbf{e}_2 \wedge \mathbf{e}_3 \wedge \mathbf{e}_4 \end{array} \right. \qquad [5.82]$$

Finally, if we calculate $\mathbf{a}_1 \wedge \mathbf{a}_2 \wedge \mathbf{a}_3$, we immediately note that it equals $\mathbf{0}$ and thus the search of the solution's existence stops at this point. Therefore, it is quite easy to observe that the linear system admits solutions if and only if $t = -2$, because we then have simultaneously:

$$\mathbf{a}_1 \wedge \mathbf{a}_2 \neq \mathbf{0} \;\; \textit{and} \;\; \mathbf{b} \wedge \mathbf{a}_1 \wedge \mathbf{a}_2 = \mathbf{0}. \qquad [5.83]$$

Thus, consider such a situation. The vector $\mathbf{b}$ can then be written as:

$$\mathbf{b} = -1\mathbf{e}_1 + 3\mathbf{e}_2 + 1\mathbf{e}_3 - 2\mathbf{e}_4. \quad [5.84]$$

Let us now find the general solutions of the linear system (written in vector form):

$$\sum_{i=1}^{i=3} x_i \mathbf{a}_i = \mathbf{b}. \quad [5.85]$$

By taking into account the relation between the vectors, we rewrite the system as:

$$x_1\mathbf{a}_1 + x_2\mathbf{a}_2 = \mathbf{b} - x_3\mathbf{a}_3, \quad [5.86]$$

which is obtained by solving:

$$x_1\mathbf{a}_1 \wedge \mathbf{a}_2 = (\mathbf{b} - x_3\mathbf{a}_3) \wedge \mathbf{a}_2, \quad x_2\mathbf{a}_2 \wedge \mathbf{a}_1 = (\mathbf{b} - x_3\mathbf{a}_3) \wedge \mathbf{a}_1. \quad [5.87]$$

In particular, the solution with respect to x_1 yields immediately:

$$x_1 \left| \begin{matrix} 2 & \mathbf{e}_1 \wedge \mathbf{e}_2 \\ 1 & \mathbf{e}_1 \wedge \mathbf{e}_3 \\ -1 & \mathbf{e}_1 \wedge \mathbf{e}_4 \\ -1 & \mathbf{e}_2 \wedge \mathbf{e}_3 \\ -1 & \mathbf{e}_2 \wedge \mathbf{e}_4 \\ -1 & \mathbf{e}_3 \wedge \mathbf{e}_4 \end{matrix} \right. = \left| \begin{matrix} -4 - 4x_3 & \mathbf{e}_1 \wedge \mathbf{e}_2 \\ -2 - 2x_3 & \mathbf{e}_1 \wedge \mathbf{e}_3 \\ 2 + 2x_3 & \mathbf{e}_1 \wedge \mathbf{e}_4 \\ 2 + 2x_3 & \mathbf{e}_2 \wedge \mathbf{e}_3 \\ 2 + 2x_3 & \mathbf{e}_2 \wedge \mathbf{e}_4 \\ 2 + 2x_3 & \mathbf{e}_3 \wedge \mathbf{e}_4 \end{matrix} \right. \quad [5.88]$$

which entails $x_1 = -2x_3 - 2$. The same procedure is carried out for the solution x_2 to find:

$$x_2 \left| \begin{matrix} -2 & \mathbf{e}_1 \wedge \mathbf{e}_2 \\ -1 & \mathbf{e}_1 \wedge \mathbf{e}_3 \\ 1 & \mathbf{e}_1 \wedge \mathbf{e}_4 \\ 1 & \mathbf{e}_2 \wedge \mathbf{e}_3 \\ 1 & \mathbf{e}_2 \wedge \mathbf{e}_4 \\ 1 & \mathbf{e}_3 \wedge \mathbf{e}_4 \end{matrix} \right. = \left| \begin{matrix} -2 + 2x_3 & \mathbf{e}_1 \wedge \mathbf{e}_2 \\ -1 + x_3 & \mathbf{e}_1 \wedge \mathbf{e}_3 \\ 1 - x_3 & \mathbf{e}_1 \wedge \mathbf{e}_4 \\ 1 - x_3 & \mathbf{e}_2 \wedge \mathbf{e}_3 \\ 1 - x_3 & \mathbf{e}_2 \wedge \mathbf{e}_4 \\ 1 - x_3 & \mathbf{e}_3 \wedge \mathbf{e}_4 \end{matrix} \right. \quad [5.89]$$

which immediately yields $x_2 = 1 - x_3$. Therefore, we can conclude by writing that the solution of the linear system:

$$x_1 \left| \begin{matrix} 1 \\ -1 \\ 0 \\ 1 \end{matrix} \right. + x_2 \left| \begin{matrix} 1 \\ 1 \\ 1 \\ 0 \end{matrix} \right. + x_3 \left| \begin{matrix} 3 \\ -1 \\ 1 \\ 2 \end{matrix} \right. = \left| \begin{matrix} -1 \\ 3 \\ 1 \\ -2 \end{matrix} \right. \quad [5.90]$$

is the set of reals x_1, x_2, x_3 such that:

$$x_1 = -2x_3 - 2, \quad x_2 = 1 - x_3, \quad x_3 \in \mathbb{R}. \qquad [5.91]$$

REMARK 5.4.– The illustrated method is quite feasible from an algorithmic point of view, as it requires only some calculations of exterior products. Of course, for vectors of a large size these calculations can be quite burdensome. In practice, if these vectors are sparse, there should be possibilities to optimize the calculations. That being said, since it is necessary to verify the dependence between vectors positively, there is no other way than calculating all the components of the exterior products. Therefore, the highest computational cost consists of the demonstration of the existence of a solution, i.e. in showing that $\mathbf{b} \wedge \mathbf{a}_{i_1} \wedge \cdots \mathbf{a}_{j_p} = \mathbf{0}$. For the rest, let us highlight the following points:

1) In order to show the independence between one $\mathbf{a}_k$ and $\mathbf{a}_{i_1}, \cdots, \mathbf{a}_{i_p}$ (supposed linearly independent), it is possible to stop as soon as one of the components of the exterior product $\mathbf{a}_{i_1} \wedge \cdots \wedge \mathbf{a}_{i_p} \wedge \mathbf{a}_k$ is non-null.

2) Once the basis $\mathbf{a}_{i_1}, \cdots, \mathbf{a}_{i_r}$ is defined in order to solve the system, the only required calculations are:

$$\forall k \in [1, r], \ \mathbf{b} \wedge (\mathbf{a}_{i_1} \wedge \cdots \wedge \mathbf{a}_{i_r} / \mathbf{a}_{i_k}), \ \ \mathbf{a}_j \wedge (\mathbf{a}_{i_1} \wedge \cdots \wedge \mathbf{a}_{i_r} / \mathbf{a}_{i_k}), \ j \notin \{i_1, \cdots, i_r\} \qquad [5.92]$$

as well as the exterior product $\mathbf{a}_{i_1} \wedge \cdots \wedge \mathbf{a}_{i_r}$.

3) However, in these calculations of exterior products, for which we shall write:

$$\begin{aligned} x_k \mathbf{a}_{i_k} \wedge (\mathbf{a}_{i_1} \wedge \cdots \wedge \mathbf{a}_{i_r} / \mathbf{a}_{i_k}) &= \mathbf{b} \wedge (\mathbf{a}_{i_1}, \wedge \cdots \wedge \mathbf{a}_{i_r} / \mathbf{a}_{i_k}) \\ &+ \sum_{j \notin \{i_1, \cdots, i_r\}} x_j \mathbf{a}_j \wedge (\mathbf{a}_{i_1} \wedge \cdots \wedge \mathbf{a}_{i_r} / \mathbf{a}_{i_k}), \end{aligned} \qquad [5.93]$$

all the calculations are actually redundant (as one can realize by performing the above calculations)! In other words, in order to obtain the expression of x_{i_k} with respect to $\mathbf{b}$ and to $x_j \mathbf{a}_j, j \notin \{i_1, \cdots, i_r\}$, it suffices to calculate a single (non-null) component of $\mathbf{a}_{i_1} \wedge \cdots \mathbf{a}_{i_r}$ and to perform the calculations of this component (and only for this one) for $\mathbf{b} \wedge (\mathbf{a}_{i_1} \wedge \cdots \wedge \mathbf{a}_{i_r} / \mathbf{a}_{i_k})$ and $\mathbf{a}_j \wedge (\mathbf{a}_{i_1} \wedge \cdots \wedge \mathbf{a}_{i_r} / \mathbf{a}_{i_k})$.

4) However, in order to avoid arithmetical inaccuracies, for better precision, it is possible to work on several components at the same time and choose the best relation among the previous ones. ■

6

Pseudo-dot Products

In this chapter, we recall some important results regarding quadratic linear spaces. Their basic purpose is to answer the following question:

QUESTION 6.1.– Suppose that we are accustomed to working with Euclidean spaces. Which results still hold within spaces that are only pseudo-Euclidean?

It is important to linger for a moment on the concept of a quadratic form that is nondegenerate and not positive definite (which thus enables the definition of a pseudo-dot product over $\mathbb{R}^4$: these are the Minkowski spaces), because it structures the whole discourse about relativity, which itself largely employs exterior formalism. Therefore, the construction of Euclidean exterior algebras (illustrated in the following chapters) would be quite incomplete if it did not immediately include the construction of pseudo-Euclidean exterior algebras.

6.1. Quadratic forms and symmetric bilinear forms

6.1.1. *Definition - correspondence*

DEFINITION 6.1.– Let $\boldsymbol{E}$ be a linear space over $\mathbb{R}$. We call bilinear form over $\boldsymbol{E}$ any map $b : \boldsymbol{E} \times \boldsymbol{E} \mapsto \mathbb{R}$ that is bilinear, i.e. linear with respect to each of its arguments. A bilinear form is called:

1) symmetric when $\forall \mathbf{x}, \mathbf{y} \in \boldsymbol{E}, b(\mathbf{x}, \mathbf{y}) = b(\mathbf{y}, \mathbf{x})$;

2) nondegenerate when $\forall \mathbf{x} \in \boldsymbol{E}, [\forall \mathbf{y} \in \boldsymbol{E},\ b(\mathbf{x}, \mathbf{y}) = 0] \Rightarrow \mathbf{x} = \mathbf{0}$;

3) positive when $\forall \mathbf{x} \in \boldsymbol{E}, b(\mathbf{x}, \mathbf{x}) \geq 0$;

4) definite when $\forall \mathbf{x} \in \boldsymbol{E},\ [b(\mathbf{x}, \mathbf{x}) = 0 \Rightarrow \mathbf{x} = 0]$.

When b is a degenerate form, an element $\mathbf{x} \neq \mathbf{0} \in \boldsymbol{E}$ satisfying $\forall \mathbf{y} \in \boldsymbol{E}, b(\mathbf{x}, \mathbf{y}) = 0$ is called a degenerate element of $\boldsymbol{E}$.

EXAMPLE 6.1.– In all that follows, we consider $\boldsymbol{E} = \mathbb{R}^2$:

1) the map $\mathbf{x}, \mathbf{y} \mapsto x_1 y_1$ is a symmetric, positive, bilinear form. However, it is degenerate since there exists an element $\mathbf{x} := \mathbf{e}_2$ such that $\forall \mathbf{y} \in \mathbb{R}^2$, we have $b(\mathbf{e}_2, \mathbf{y}) = 0$;

2) the map $\mathbf{x}, \mathbf{y} \mapsto x_1 y_1 - x_2 y_2$ is clearly a symmetric bilinear form. It is not definite since:

$$b(\mathbf{e}_1 + \mathbf{e}_2, \mathbf{e}_1 + \mathbf{e}_2) = 1.1 - 1.1 = 0, \tag{6.1}$$

therefore, it is possible to find $\mathbf{x} \neq \mathbf{0}$ such that $b(\mathbf{x}, \mathbf{x}) = 0$. On the other hand, this form is nondegenerate. In fact, suppose that we have found $\mathbf{x} = x_1 \mathbf{e}_1 + x_2 \mathbf{e}_2$ such that:

$$\forall \mathbf{y} \in \mathbb{R}^2, \; b(\mathbf{x}, \mathbf{y}) = 0, \tag{6.2}$$

in that case, by choosing at first $\mathbf{y} = x_1 \mathbf{e}_1$ we obtain $x_1^2 = 0$ (and thus $x_1 = 0$), then, by choosing $\mathbf{y} = x_2 \mathbf{e}_2$ we obtain $x_2^2 = 0$ (and thus $x_2 = 0$) which yields indeed $\mathbf{x} = \mathbf{0}$;

3) the map $\mathbf{x}, \mathbf{y} \mapsto x_1 y_1 + x_2 y_2$, the common dot product of $\mathbb{R}^2$, is a symmetric positive-definite nondegenerate form. The positive-definite characteristics are evident since:

$$b(\mathbf{x}, \mathbf{x}) = x_1^2 + x_2^2 \geq 0 \;\; et \;\; b(\mathbf{x}, \mathbf{x}) = 0 \Leftrightarrow x_1^2 + x_2^2 = 0 \Leftrightarrow x_1 = x_2 = 0 \Leftrightarrow \mathbf{x} = \mathbf{0}, \tag{6.3}$$

while the nondegenerate aspect can be shown exactly as in the previous example. ■

We recall the usual vocabulary:

DEFINITION 6.2.– Let $\boldsymbol{E}$ be a linear space over $\mathbb{R}$. We call:

1) dot product over $\boldsymbol{E}$ any symmetric, positive-definite bilinear form;

2) pseudo-dot product over $\boldsymbol{E}$ any symmetric and nondegenerate bilinear form.

A finite dimensional linear space equipped with a dot product (respectively, pseudo-dot product) is called a Euclidean space (respectively, pseudo-Euclidean).

At this point, let us address the notion of a quadratic form:

DEFINITION 6.3.– Let $\boldsymbol{E}$ be a linear space over $\mathbb{R}$. We call quadratic form over $\boldsymbol{E}$ any map $q : \boldsymbol{E} \mapsto \mathbb{R}$ such that:

1) for all $\lambda \in \mathbb{R}$ and for all $\mathbf{x} \in \boldsymbol{E}$, we have $q(\lambda \mathbf{x}) = \lambda^2 q(\mathbf{x})$;

2) the map $\mathbf{x}, \mathbf{y} \mapsto \frac{1}{2}(q(\mathbf{x}+\mathbf{y}) - q(\mathbf{x}) - q(\mathbf{y}))$ is a symmetric bilinear form.

A quadratic form is called:

1) positive when $\forall \mathbf{x} \in \boldsymbol{E},\ q(\mathbf{x}) \geq 0$;

2) definite when $\forall \mathbf{x} \in \boldsymbol{E}, [q(\mathbf{x}) = 0 \Rightarrow \mathbf{x} = \mathbf{0}]$.

An element $\mathbf{x}$ of $\boldsymbol{E}$ that simultaneously satisfies $q(\mathbf{x}) = 0$ and $\mathbf{x} \neq \mathbf{0}$ is called an isotropic vector. We call anisotropic vector a vector that verifies $q(\mathbf{x}) \neq 0$.

In particular, an anisotropic vector is never null, while the reciprocal statement is true only if the form q is definite. The following theorem highlights the relation between quadratic forms and symmetric bilinear forms:

THEOREM 6.1.– Let $q : \boldsymbol{E} \mapsto \mathbb{R}$ be a quadratic form. Then, there exists a unique symmetric bilinear form $b : \boldsymbol{E} \times \boldsymbol{E} \mapsto \mathbb{R}$ such that:

$$\forall \mathbf{x} \in \boldsymbol{E},\quad q(\mathbf{x}) = b(\mathbf{x}, \mathbf{x}). \qquad [6.4]$$

PROOF.– First of all, it is clear that the set of symmetric bilinear forms over $\boldsymbol{E}$, denoted by $S(\boldsymbol{E})$, is a linear space, and the same is true for the set of quadratic forms over $\boldsymbol{E}$, denoted by $Q(\boldsymbol{E})$. Let b be a bilinear form. It is possible to associate it with a map $L(b)(\cdot) : \boldsymbol{E} \mapsto \mathbb{R}$ according to the expression:

$$\forall \mathbf{x} \in \boldsymbol{E},\quad L(b)(\mathbf{x}) := b(\mathbf{x}, \mathbf{x}). \qquad [6.5]$$

Therefore, it is quite evident that $L(b)$ is a quadratic form. On the other hand, in virtue of its very construction, the map $L(\cdot) : S(\boldsymbol{E}) \mapsto Q(\boldsymbol{E})$ is linear. We shall show that this map is actually a linear space isomorphism. According to the point 2 of definition 6.3, the map L is indeed surjective from $S(\boldsymbol{E})$ to $Q(\boldsymbol{E})$. In fact, consider $q \in Q(\boldsymbol{E})$. Then, the map

$$b(\mathbf{x}, \mathbf{y}) := \frac{1}{2}(q(\mathbf{x}+\mathbf{y}) - q(\mathbf{x}) - q(\mathbf{y})) \qquad [6.6]$$

is an element of $S(\boldsymbol{E})$. Clearly, we have:

$$L(b)(\mathbf{x}) = b(\mathbf{x},\mathbf{x}) = \frac{1}{2}(q(2\mathbf{x}) - q(\mathbf{x}) - q(\mathbf{x})), \qquad [6.7]$$

in virtue of the axiom 1 from definition 6.3, we know that $q(2\mathbf{x}) = 4q(\mathbf{x})$ and we deduce that:

$$L(b)(\mathbf{x}) = \frac{1}{2}(4q(\mathbf{x}) - 2q(\mathbf{x})) = q(\mathbf{x}), \qquad [6.8]$$

hence the surjectivity of $L(\cdot)$. Let us now consider the injectivity. For this purpose, by fixing b within $S(\boldsymbol{E})$, let us note $q = L(b)$. Clearly, we have:

$$\begin{aligned} &\frac{1}{2}(q(\mathbf{x}+\mathbf{y}) - q(\mathbf{x}) - q(\mathbf{y})) \\ &= \frac{1}{2}(b(\mathbf{x}+\mathbf{y},\mathbf{x}+\mathbf{y}) - b(\mathbf{x},\mathbf{x}) - b(\mathbf{y},\mathbf{y})) = b(\mathbf{x},\mathbf{y}). \end{aligned} \qquad [6.9]$$

In particular, if $q = L(b) = 0$ it entails immediately that indeed $b = 0$, which demonstrates the injectivity of L since L is linear.

6.1.2. *Sign of the forms: on the possibility of restricting the forms to subspaces*

An important question regarding bilinear or quadratic forms concerns the capability of maintaining their fundamental properties when they are restricted to some linear subspaces.

DEFINITION 6.4.– Given a bilinear form $b : \boldsymbol{E} \times \boldsymbol{E} \mapsto \mathbb{R}$, it is always possible to consider $b_{\boldsymbol{F}}$, a form defined as:

$$\forall \mathbf{y},\mathbf{x} \in \boldsymbol{F}, \ \ b_{\boldsymbol{F}}(\mathbf{x},\mathbf{y}) := b(\mathbf{x},\mathbf{y}). \qquad [6.10]$$

We call $b_{\boldsymbol{F}}$ the restriction of b to $\boldsymbol{F}$.

QUESTION 6.2.– However, a question arises whether the "good" properties of b over $\boldsymbol{E}$ still exist for $b_{\boldsymbol{F}}$ over $\boldsymbol{F}$.

From a formal point of view, some doubts are legitimate to the extent that the fundamental properties of b are stated with formulas of the form: $\forall \mathbf{x} \in \boldsymbol{E} \cdots$ or $\exists \mathbf{x} \in \boldsymbol{E} \cdots$, and it is not evident at all that these formulas can be replaced ipso facto by $\forall \mathbf{x} \in \boldsymbol{F}$

or $\exists \mathbf{x} \in \boldsymbol{F}$, which would enable an inference over $\boldsymbol{F}$ of the properties of b over $\boldsymbol{E}$. The following vocabulary is introduced:

DEFINITION 6.5.– Let $(\boldsymbol{E}, b)$ be a pseudo-Euclidean space and $\boldsymbol{F}$ a subspace of $\boldsymbol{E}$. We say that $\boldsymbol{F}$ is nondegenerate (respectively, definite) when the restriction $b_{\boldsymbol{F}}$ is nondegenerate (respectively, definite).

PROPOSITION 6.1.– Let $b : \boldsymbol{E} \times \boldsymbol{E} \mapsto \mathbb{R}$ be a nondegenerate symmetric bilinear form over $\boldsymbol{E}$. Let $\boldsymbol{F} \subset \boldsymbol{E}$, $\boldsymbol{F} \neq \boldsymbol{E}$ a subspace of $\boldsymbol{E}$. In that case, the form restricted to $b_{\boldsymbol{F}}$ can be degenerated.

PROOF.– The demonstration will be illustrated later, using the orthogonality vocabulary (see corollary 6.3).

In practice, the sign of the form will indicate whether it is possible to infer all the good properties of the form b over $\boldsymbol{E}$ within the subspaces $\boldsymbol{F}$.

PROPOSITION 6.2.– Let $b : \boldsymbol{E} \times \boldsymbol{E} \mapsto \mathbb{R}$ be a symmetric bilinear form. In that case, we have the following equivalence:

1) the form b is positive definite;

2) the form b is nondegenerate positive.

PROOF.– Suppose that b is nondegenerate positive and suppose that $b(\mathbf{x}, \mathbf{x}) = 0$. In that case, for all $\lambda, \mathbf{y}$, we have:

$$b(\mathbf{x} + \lambda \mathbf{y}, \mathbf{x} + \lambda \mathbf{y}) = \lambda^2 b(\mathbf{y}, \mathbf{y}) + 2\lambda b(\mathbf{x}, \mathbf{y}), \quad [6.11]$$

the left-hand side is always positive or null, independently of the values of λ and $\mathbf{y}$. Suppose that there exists $\mathbf{y}$ such that $b(\mathbf{x}, \mathbf{y}) \neq 0$. In that case, for λ sufficiently small and with the correct sign, it may be possible to obtain $\lambda^2 b(\mathbf{y}, \mathbf{y}) + 2\lambda b(\mathbf{x}, \mathbf{y}) < 0$: this is, however, impossible since the left-hand side is positive. Thus, we deduce that:

$$\forall \mathbf{y} \in \boldsymbol{E}, \ b(\mathbf{x}, \mathbf{y}) = 0, \quad [6.12]$$

which entails that $\mathbf{x} = \mathbf{0}$ since the form b is nondegenerate. Thus, we indeed have that nondegenerate positivity implies definite positivity. Let us consider the reciprocal statement. Suppose that b is definite (positive) and suppose that:

$$\forall \mathbf{y}, \ b(\mathbf{x}, \mathbf{y}) = 0, \quad [6.13]$$

then, in particular for $\mathbf{y} = \mathbf{x}$, we have $b(\mathbf{x}, \mathbf{x}) = 0$ which entails $\mathbf{x} = 0$ since b is definite.

The conclusion of the demonstration does not require any hypothesis about the sign of the form. In fact, evidently, we have the following proposition:

PROPOSITION 6.3.– Let b be a definite symmetric bilinear form. Then, it is nondegenerate.

PROOF.– See above.

6.1.3. *Representation via nondegenerate forms*

DEFINITION 6.6.– Let $\boldsymbol{E}$ be a space with a finite dimension n equipped with a bilinear (not necessarily symmetric) form b. For any basis $\mathcal{B} := (\mathbf{e}_1, \cdots, \mathbf{e}_n)$ of $\boldsymbol{E}$, we call representative matrix of b with respect to $\mathcal{B}$ the matrix $\mathbb{B}$ such that:

$$\forall (i,j) \in [1,n]^2, \ \mathbb{B}_{i,j} = b(\mathbf{e}_i, \mathbf{e}_j). \tag{6.14}$$

Consequently, we have a rather simple theorem:

PROPOSITION 6.4.– Let $\boldsymbol{E}$ be a space of a finite dimension n equipped with a bilinear form b. Then, $\boldsymbol{E}$ is pseudo-Euclidean (respectively, Euclidean) if and only if there exists a basis $\mathcal{B}$ of $\boldsymbol{E}$ such that the matrix $\mathbb{B}$ of b in $\mathcal{B}$ is symmetric and invertible (respectively, positive-definite symmetric). In such case, for any basis, the representative matrix is symmetric and invertible (respectively, positive-definite symmetric).

PROOF.– Suppose that b is pseudo-Euclidean. Let us consider some basis of $\boldsymbol{E}$. Clearly, we have:

$$\mathbb{B}_{i,j} = b(\mathbf{e}_i, \mathbf{e}_j) = b(\mathbf{e}_j, \mathbf{e}_i) = \mathbb{B}_{j,i}, \tag{6.15}$$

and therefore the matrix $\mathbb{B}$ is indeed symmetric. On the other hand, suppose that for a column vector $[x_1, \cdots, x_n]^T$, we have:

$$\forall i \in [1,n], \ \sum_{j=1}^{j=n} \mathbb{B}_{ij} x_j = 0, \tag{6.16}$$

then we deduce that for any row vector $[y_1, \cdots, y_n]$, we have:

$$\sum_{ij} y_j \mathbb{B}_{ij} x_j = 0, \tag{6.17}$$

and thus by interpreting the matrix $\mathbb{B}$ and by considering the vectors $\mathbf{y} = \sum_j y_j \mathbf{e}_j$ and $\mathbf{x} = \sum_i x_i \mathbf{e}_i$, we deduce that:

$$\forall \mathbf{y} \in \boldsymbol{E}, b(\mathbf{y}, \mathbf{x}) = 0, \qquad [6.18]$$

which implies that $\mathbf{x} = 0$ and thus that $\forall i \in [1, n]$, $x_i = 0$ since the form b is nondegenerate. The matrix $\mathbb{B}$ is thus invertible. Vice versa, suppose that there exists a basis such that the matrix $\mathbb{B}$ is symmetric and invertible. In that case, it can be shown that actually $b(\mathbf{e}_i, \mathbf{e}_j) = b(\mathbf{e}_j, \mathbf{e}_i)$ for all the vectors of the basis and therefore, in virtue of bilinearity, also for all the vectors $\mathbf{x}, \mathbf{y}$. Now, if $\mathbb{B}$ is invertible, we have:

$$\forall \mathbf{y} \in \boldsymbol{E}, b(\mathbf{x}, \mathbf{y}) = 0 \Rightarrow \forall i \in [1, n], \sum_{j=1}^{j=n} \mathbb{B}_{ij} x_j = 0, \qquad [6.19]$$

and thus $\mathbf{x} = \sum_i x_i \mathbf{e}_i = \mathbf{0}$ since $\mathbb{B}$ is invertible. Then for any basis, since b is nondegenerate and definite, the matrix $\mathbb{B}$, representing b within $\mathcal{B}$, becomes symmetric and invertible.

The following corollary is immediately deduced:

COROLLARY 6.1.– Let $(\boldsymbol{E}, b)$ be a pseudo-Euclidean space and $\mathbf{e}_i, i \in [1, n]$ a pseudo-orthogonal basis. Then, we have $\forall i \in [1, n]$, $q(\mathbf{e}_i) \neq 0$.

PROOF.– The matrix $\mathbb{B}$ representing b is diagonal over the basis of $\mathbf{e}_i, i \in [1, n]$, with diagonal values equal to $q(\mathbf{e}_i)$. If one of these values was null, the matrix $\mathbb{B}$ would not be invertible.

The Riesz representation theorem, elementary in the case of finite dimensions, will become crucially important in the following work on exterior algebra:

THEOREM 6.2.– [Riesz representation] Let $b : \boldsymbol{E} \times \boldsymbol{E} \mapsto \mathbb{R}$ be a nondegenerate bilinear (not necessarily symmetric) form over a finite dimensional linear space $\boldsymbol{E}$. Then, for all linear form $l \in \boldsymbol{E}^*$, there exists a unique vector $\mathbf{x}$ of $\boldsymbol{E}$ such that:

$$\forall \mathbf{y} \in \boldsymbol{E}, \quad b(\mathbf{x}, \mathbf{y}) = l(\mathbf{y}). \qquad [6.20]$$

PROOF.– Consider the map:

$$\forall \mathbf{x} \in \boldsymbol{E} \mapsto b(\mathbf{x}, \cdot) \in \boldsymbol{E}^*, \qquad [6.21]$$

clearly, this map is linear. Furthermore, since b is nondegenerate, it is straightforward that this map is also injective. We know as well that $\boldsymbol{E}$ and $\boldsymbol{E}^*$ have the same dimension: as a result, injectivity is equivalent to bijectivity.

REMARK 6.1.– Of course, the purpose of the Riesz theorem is to associate an element $\mathbf{x}$ with all linear form, and to do this in a unique way. Actually, in practice, linear forms appear primitively in several situations in physics or in other modeling domains. For instance, it is possible to define the notion of force in a dual way as being a linear form operating over velocity vectors. The vector associated via the Riesz theorem can then be considered as the force vector. This kind of consideration is, for example, the foundation of concepts of virtual work. ■

An immediate application of the Riesz representation theorem consists of the definition of pseudo-adjoint:

DEFINITION 6.7.– Let $u : \boldsymbol{E}_1 \mapsto \boldsymbol{E}_2$ be a linear operator between two pseudo-Euclidean spaces $(\boldsymbol{E}_1, b_1)$ and $(\boldsymbol{E}_2, b_2)$. Then, there exists a unique linear operator u^* between $\boldsymbol{E}_2 \mapsto \boldsymbol{E}_1$ such that:

$$\forall \mathbf{x} \in \boldsymbol{E}_1, \forall \mathbf{y} \in \boldsymbol{E}_2, \;\; b_2\left(u\left(\mathbf{x}\right), \mathbf{y}\right) = b_1\left(\mathbf{x}, u^*\left(\mathbf{y}\right)\right), \qquad [6.22]$$

the operator u^* is called pseudo-adjoint of u (for the forms b_1, b_2).

PROOF.– For a fixed $\mathbf{y} \in \boldsymbol{E}_2$, the operator $b_2\left(u\left(\cdot\right), \mathbf{y}\right)$ defines a linear form for $\mathbf{x} \in \boldsymbol{E}_1$. In virtue of the Riesz representation theorem, there exists a unique element of $\boldsymbol{E}_1$, denoted by $\mathbf{Y} := u^*\left(\mathbf{y}\right)$ such that:

$$\forall \mathbf{x} \in \boldsymbol{E}_1, \;\; b_1\left(\mathbf{Y}, \mathbf{x}\right) = b_2\left(u\left(\mathbf{x}\right), \mathbf{y}\right). \qquad [6.23]$$

Since b_1 is nondegenerate, it can easily be seen that the operator $\mathbf{y} \mapsto u^*\left(\mathbf{y}\right)$ is linear, because:

$$\forall \mathbf{x} \in \boldsymbol{E}_1, \forall \mathbf{y}, \mathbf{y}' \in \boldsymbol{E}_2, \forall \lambda \in \mathbb{R}, \;\; b_1\left(u^*\left(\mathbf{y} + \lambda \mathbf{y}'\right), \mathbf{x}\right) = b_2\left(\mathbf{y} + \lambda \mathbf{y}', u\left(\mathbf{x}\right)\right) \qquad [6.24]$$

by linearity of b_2, the right-hand side is equal to:

$$b_2\left(\mathbf{y}, u\left(\mathbf{x}\right)\right) + \lambda b_2\left(\mathbf{y}', u\left(\mathbf{x}\right)\right) = b_1\left(u^*\left(\mathbf{y}\right), \mathbf{x}\right) + \lambda b_1\left(u^*\left(\mathbf{y}'\right), \mathbf{x}\right), \qquad [6.25]$$

hence, we deduce that:

$$\forall \mathbf{x} \in \boldsymbol{E}_1, \forall \mathbf{y}, \mathbf{y}' \in \boldsymbol{E}_2, \forall \lambda \in \mathbb{R},$$
$$b_1\left(u^*\left(\mathbf{y} + \lambda \mathbf{y}'\right), \mathbf{x}\right) = b_1\left(u^*\left(\mathbf{y}\right) + \lambda u^*\left(\mathbf{y}'\right), \mathbf{x}\right), \qquad [6.26]$$

and since the form b_1 is nondegenerate, this entails that:

$$\forall \mathbf{y}, \mathbf{y}' \in \boldsymbol{E}_2, \forall \lambda \in \mathbb{R}, \;\; u^*\left(\mathbf{y} + \lambda \mathbf{y}'\right) = u^*\left(\mathbf{y}\right) + \lambda u^*\left(\mathbf{y}'\right). \qquad [6.27]$$

Some consequent definitions can be introduced:

DEFINITION 6.8.– Let $(\boldsymbol{E}, b)$ be a pseudo-Euclidean space and u an endomorphism of $\boldsymbol{E}$. We say that:

1) the map u is pseudo-autoadjoint (for b), when $u = u^*$;

2) the map u is pseudo-orthogonal (for b), when $b(u(\mathbf{x}), u(\mathbf{y})) = b(\mathbf{x}, \mathbf{y})$. This means that u is invertible and it satisfies $u^{-1} = u^*$.

REMARK 6.2.– In the context of physics, the study of isometries of the spacetime, which is a space equipped with a pseudo-dot product, is extremely important; therefore, it is absolutely legitimate to introduce this notion for pseudo-Euclidean spaces and not exclusively for Euclidean spaces. ■

6.2. Orthogonality

6.2.1. *Definition - elementary properties*

DEFINITION 6.9.– Let $(\mathbf{E}, \mathbf{b})$ be a pseudo-Euclidean space. For all subset $A \subset \boldsymbol{E}$, we call orthogonal of A in $\boldsymbol{E}$, denoted by $A^\perp$, the following set:

$$A^\perp = \{\mathbf{x} \in \boldsymbol{E},\ \forall \mathbf{y} \in A,\ b(\mathbf{x}, \mathbf{y}) = 0\}. \quad [6.28]$$

Clearly, by employing the bilinearity of the pseudo-dot product, the set $A^\perp$ is always a linear subspace of $\boldsymbol{E}$. The following theorem will be useful for characterizing pseudo-dot products:

PROPOSITION 6.5.– Let $\boldsymbol{F}$ be a linear subspace of the pseudo-Euclidean space $(\boldsymbol{E}, b)$ with a dimension n. Then, we have:

$$\dim(\boldsymbol{F}^\perp) = n - \dim(\boldsymbol{F}). \quad [6.29]$$

PROOF.– Let $\mathbf{e}_1, \cdots, \mathbf{e}_p$ be a basis of $\boldsymbol{F}$ which we complete with $\mathbf{e}_{p+1}, \cdots, \mathbf{e}_n$ to obtain a basis of $\boldsymbol{E}$. Let $\mathbf{x} = \sum_{i=1}^{i=n} x_i \mathbf{e}_i$ be an element (*a priori*, whichever one) of $\boldsymbol{E}$. We have:

$$\mathbf{x} \in \boldsymbol{F}^\perp \Leftrightarrow \forall i \in [1, p],\ b(\mathbf{e}_i, \mathbf{x}) = 0 \Leftrightarrow \forall i \in [1, p],\ \sum_{j=1}^{j=n} b(\mathbf{e}_i, \mathbf{e}_j)\, x_j = 0. \quad [6.30]$$

Since the family $\mathbf{e}_1, \cdots, \mathbf{e}_n$ is a basis and the form b is nondegenerate, the matrix of $b(\mathbf{e}_i, \mathbf{e}_j), i, j \in [1, n]$ is a matrix with a rank n (see proposition 6.4). As a result, the matrix $b(\mathbf{e}_i, \mathbf{e}_j), i \in [1, p], j \in [1, n]$ has a rank p. The set of $\mathbf{x}$ such that:

$$\forall i \in [1, p], \sum_{j=1}^{j=n} b(\mathbf{e}_i, \mathbf{e}_j) x_j = 0 \quad [6.31]$$

is therefore a linear subspace of dimension $n - p$, which is thus the dimension of $F^{\perp}$.

Hence, we deduce the following corollary:

COROLLARY 6.2.– Let F be a linear subspace of the pseudo-Euclidean space (E, b). Then, we have:

$$\left(F^{\perp}\right)^{\perp} = F. \quad [6.32]$$

PROOF.– According to the definition of orthogonal subspace 6.9, it is clear that the inclusion is always true:

$$F \subset \left(F^{\perp}\right)^{\perp}, \quad [6.33]$$

we apply then the above dimension formula (proposition 6.5) and we obtain:

$$\dim\left(\left(F^{\perp}\right)^{\perp}\right) = \dim(F), \quad [6.34]$$

since F and $\left(F^{\perp}\right)^{\perp}$ are two linear subspaces (included one within another), we finally deduce that they are equal.

Let us continue with a non-trivial structural theorem that provides a partial answer to the questions regarding the restriction of nondegenerate forms:

THEOREM 6.3.– Let F be a linear subspace of E. Then, F is nondegenerate (see definition 6.5) if and only if $F^{\perp}$ is nondegenerate.

PROOF.– To the extent that $F = \left(F^{\perp}\right)^{\perp}$, it is enough to show the implication (rather than the equivalence). Suppose then that F is nondegenerate. This means that the restriction b_F of b to the space F is itself nondegenerate. Imagine now that $F^{\perp}$ is degenerate. Two cases are possible:

1) Suppose that $F \cap F^{\perp} = \{0\}$. In that case, we have $E = F \oplus^{\perp} F^{\perp}$ and every element $\mathbf{x}$ of E is written in a unique way:

$$\mathbf{x} = \mathbf{x}_F + \mathbf{x}_F^{\perp}, \quad \mathbf{x}_F \in F, \quad \mathbf{x}_F^{\perp} \in F^{\perp}. \qquad [6.35]$$

Then, if the restriction of b to $F^{\perp}$ is degenerate, let us denote by $\mathbf{y} \in F^{\perp}$ a non-null element of $F^{\perp}$ such that $b_{F^{\perp}}(\mathbf{y},\cdot) = 0$. In that case, it is clear that this $\mathbf{y}$ satisfies:

$$\forall \mathbf{x} \in E, \; b(\mathbf{x},\mathbf{y}) = b\left(\mathbf{x}_E + \mathbf{x}_E^{\perp},\mathbf{y}\right) = b(\mathbf{x}_E,\mathbf{y}) + b_{F^{\perp}}\left(\mathbf{x}_E^{\perp},\mathbf{y}\right) = 0, \qquad [6.36]$$

as a result, b would be degenerate: impossible.

2) Suppose now that $F \cap F^{\perp} \neq \{\mathbf{0}\}$. Suppose also that $\mathbf{y} \neq 0$ is such that $b_{F^{\perp}}(\mathbf{y},\cdot) = 0$. In that case, we have:

$$\forall \mathbf{x} \in F^{\perp}, \quad b(\mathbf{x},\mathbf{y}) = b_{F^{\perp}}(\mathbf{x},\mathbf{y}) = 0, \qquad [6.37]$$

this proves that $\mathbf{y} \in \left(F^{\perp}\right)^{\perp}$ and thus that $\mathbf{y} \in F$ in virtue of corollary 6.2. Therefore, we have $\mathbf{y} \in F \cap F^{\perp}$. But then we have:

$$\forall \mathbf{x} \in E, \quad b(\mathbf{x},\mathbf{y}) = 0 = b_F(\mathbf{x},\mathbf{y}), \qquad [6.38]$$

which shows that b_F is degenerate: impossible.

As a result, we have indeed that b_F nondegenerate implies $b_{F^{\perp}}$ nondegenerate, and vice versa, since $\left(F^{\perp}\right)^{\perp} = F$ (see corollary 6.2).

Hence, from this demonstration we obtain a useful corollary:

COROLLARY 6.3.– Let F be a linear subspace of the pseudo-Euclidean space (E, b). Then, we have the following equivalence:

$$b_F \textit{ is nondegenerate} \Leftrightarrow F \cap F^{\perp} = \{0\} \Leftrightarrow b_{F^{\perp}} \textit{ is nondegenerate.} \qquad [6.39]$$

PROOF.– In the previous demonstration, we have seen that the implication $F \cap F^{\perp} = \{0\}$ entails that neither F nor $F^{\perp}$ can be degenerated, otherwise b would be degenerate. Vice versa, if we suppose that F is nondegenerate, then it is impossible that $F \cap F^{\perp}$ admits a non-null element, because if $\mathbf{y} \neq \mathbf{0} \in F \cap F^{\perp}$, $b_F(\mathbf{y},\cdot)$ is null and thus F is degenerate.

6.2.2. *Existence and construction of orthonormal bases*

Let us start with a definition:

DEFINITION 6.10.– Let $(\boldsymbol{E}, b)$ be a pseudo-Euclidean space. We call pseudo-orthonormal family of $\boldsymbol{E}, b$ all family $\mathbf{e}_i, i \in [1, p]$ such that:

$$\forall i \neq j \in [1,p] \;\; b(\mathbf{e}_i, \mathbf{e}_j) = 0, \;\; \forall i \in [1,p], \; |q(\mathbf{e}_i)| = 1 \qquad [6.40]$$

a pseudo-orthonormal family which is a basis is called a pseudo-orthonormal basis.

Let us continue with an important lemma:

LEMMA 6.1.– Let $(\boldsymbol{E}, b)$ be a pseudo-Euclidean space. There exists $\mathbf{e}_1 \in \boldsymbol{E}$ such that $q(\mathbf{e}_1) \neq 0$.

PROOF.– Suppose, on the contrary, that we have $\forall \mathbf{z} \in \boldsymbol{E},\ q(\mathbf{z}) = 0$. In that case, we would have:

$$\forall \mathbf{x}, \mathbf{y} \in \boldsymbol{E}, \;\; \frac{1}{2}\left(q(\mathbf{x}+\mathbf{y}) - q(\mathbf{x}) - q(\mathbf{y})\right) = 0, \qquad [6.41]$$

the form b would thus be null. More specifically, it would be degenerate.

This leads directly to the existence of orthonormal bases.

THEOREM 6.4.– Let $(\boldsymbol{E}, b)$ be a pseudo-Euclidean space. Then there exists a pseudo-orthonormal basis for the pseudo-dot product.

PROOF.– The demonstration applies an induction over the dimension of $\boldsymbol{E}$. Suppose that $\boldsymbol{E}$ has a dimension equal to 1. According to the previous lemma 6.1, there exists $\mathbf{x} \in \boldsymbol{E}$ such that $q(\mathbf{x}) \neq 0$. By setting $\mathbf{e}_1 = \mathbf{x}/|q(\mathbf{x})|^{1/2}$, the vector $\mathbf{e}_1$ is a pseudo-orthonormal basis of $\boldsymbol{E}$. Suppose now that for $n > 1$, all pseudo-Euclidean spaces with a dimension lower than or equal to $n-1$ admit a pseudo-orthonormal basis. Again, according to lemma 6.1, there exists a vector $\mathbf{x}$ such that $q(\mathbf{x}) \neq 0$. Let us then set $\mathbf{e}_1 = \mathbf{x}/|q(\mathbf{x})|^{1/2}$ and consider $\boldsymbol{F} = vec(\mathbf{e}_1)$ the subspace generated by $\mathbf{e}_1$. Clearly, the form b restricted to $\boldsymbol{F}$ is nondegenerate, since for $\mathbf{x} = \alpha\mathbf{e}_1, \mathbf{y} = \beta\mathbf{e}_1 \in \boldsymbol{F}$ we have:

$$b(\mathbf{x}, \mathbf{y}) = \alpha\beta b(\mathbf{e}_1, \mathbf{e}_1) = \alpha\beta q(\mathbf{e}_1), \qquad [6.42]$$

which is indeed nondegenerate. Therefore, $\boldsymbol{F}$ is nondegenerate. However, according to corollary 6.3, this implies that $\boldsymbol{F}^{\perp}$ is nondegenerate as well. Thus, $(\mathbf{F}, b_{\mathbf{F}})$ is a pseudo-Euclidean space. Besides, according to the formula of dimension (see proposition 6.5), we have that $\dim\left(\boldsymbol{F}^{\perp}\right) = n-1$. By applying the inductive hypothesis to $\boldsymbol{F}^{\perp}$: therefore, we know that there exists a pseudo-orthonormal basis $\mathbf{e}_2, \cdots, \mathbf{e}_n$. Together with the vector $\mathbf{e}_1$, we have a pseudo-orthonormal basis of $\boldsymbol{E}$.

COROLLARY 6.4.– Let $\mathbf{e}_1, \cdots, \mathbf{e}_p$ be a pseudo-orthonormal family of $\boldsymbol{E}$. Then, it can be completed by a pseudo-orthonormal family $\mathbf{e}_{p+1}, \cdots, \mathbf{e}_n$ in order to create a pseudo-orthonormal basis of $\boldsymbol{E}$.

PROOF.– Consider the space $\boldsymbol{F}$ generated by the family $\mathbf{e}_1, \cdots, \mathbf{e}_p$. Since the family is pseudo-orthonormal, the space $\boldsymbol{F}$ is nondegenerate (in fact, the matrix $\mathbb{B}$ that represents the form b within the basis of $\mathbf{e}_i, i \in [1,p]$ is diagonal with terms equal to ± 1, and therefore such a matrix is invertible and we can apply proposition 6.4). As a result, $\boldsymbol{F}^{\perp}$ is nondegenerate and we can construct a pseudo-orthonormal basis $\mathbf{e}_{p+1}, \cdots, \mathbf{e}_n$ (see theorem 6.6) of $(\boldsymbol{F}, b_{\boldsymbol{F}^{\perp}})$. Finally, the basis $\mathbf{e}_1, \cdots, \mathbf{e}_n$ is a pseudo-orthonormal basis of $(\boldsymbol{E}, b)$.

In order to construct orthonormal bases, the theorem of Gram-Schmidt with pivoting is the most convenient one:

THEOREM 6.5.– Let $(\boldsymbol{E}, b)$ be a pseudo-Euclidean space and $\mathbf{x}_1, \cdots, \mathbf{x}_p$ a linearly independent family of $\boldsymbol{E}$ such that the space $\boldsymbol{F}$ generated by this family is nondegenerate. In that case, there exists $\mathbf{e}_1, \cdots, \mathbf{e}_p$ a pseudo-orthonormal family of $\boldsymbol{E}$ and a permutation $\sigma \in S_p$ (which can be interpreted as pivoting) such that:

$$\forall k \in [1,p],\ vec\left(\mathbf{x}_{\sigma(1)}, \cdots, \mathbf{x}_{\sigma(k)}\right) = vec\left(\mathbf{e}_1, \cdots, \mathbf{e}_k\right). \qquad [6.43]$$

REMARK 6.3.– In the Euclidean case, pivoting can be avoided, i.e. the theorem of Gram-Schmidt is true for $\sigma = Id$. In the pseudo-Euclidean case, pivoting is essential, since otherwise we might encounter an isotropic vector, which never happens in the Euclidean case. In practice, pivoting is always advisable in the Euclidean case, even if it can become costly, and after all the construction of orthonormal bases is roughly the same. ■

PROOF.– We choose i such that:

$$|q(\mathbf{x}_i)| = \max_{j \in [1,n]} |q(\mathbf{x}_j)|, \qquad [6.44]$$

it is straightforward that $q(\mathbf{x}_i) \neq 0$, otherwise the space $\boldsymbol{F}$ would be degenerate. Let us then consider a transposition τ_1 that permutes 1 and i (maintaining all the other indexes) and we define:

$$\mathbf{e}_1 := \frac{\mathbf{x}_{\tau_1(1)}}{\sqrt{\left|q\left(\mathbf{x}_{\tau_1(1)}\right)\right|}}. \qquad [6.45]$$

Evidently, the space $vec(\mathbf{e}_1)$ is nondegenerate and therefore its orthogonal (within $\boldsymbol{F}$) is also nondegenerate. Consider now the family:

$$\mathbf{x}_{\tau_1(2)}, \cdots, \mathbf{x}_{\tau_1(n)} \qquad [6.46]$$

$$\forall i \in [2,n], \ \mathbf{x}_{i,1} := \mathbf{x}_{\tau_1(i)} - q(\mathbf{e}_1)\, b\left(\mathbf{x}_{\tau_1(i)}, \mathbf{e}_1\right) \mathbf{e}_1 \qquad [6.47]$$

In such case, the family of $\mathbf{x}_{i,1}, i \in [2,n]$ is a linearly independent family of $[vec(\mathbf{e}_1)]^{\perp}$. In fact, first of all, it is evident that each $\mathbf{x}_{i,1}$ is indeed orthogonal to $\mathbf{e}_1$, because:

$$\forall i \in [2,n], \ b(\mathbf{x}_{i,1}, \mathbf{e}_1) = b\left(\mathbf{x}_{\tau_1(i)}, \mathbf{e}_1\right) - q(\mathbf{e}_1)\, b\left(\mathbf{x}_{\tau_1(i)}, \mathbf{e}_1\right) b(\mathbf{e}_1, \mathbf{e}_1) = 0, \quad [6.48]$$

since $q(\mathbf{e}_1)\, b(\mathbf{e}_1, \mathbf{e}_1) = q(\mathbf{e}_1)^2 = 1$. Furthermore, the family is indeed linearly independent to the extent that:

$$\sum_{i=2}^{i=p} \lambda_i \mathbf{x}_{i,1} = \mathbf{0} \Leftrightarrow \sum_{i=2}^{i=p} \lambda_i \left(\mathbf{x}_{\tau_1(i)} - q(\mathbf{e}_1)\, b\left(\mathbf{x}_{\tau_1(i)}, \mathbf{e}_1\right) \mathbf{e}_1\right) = \mathbf{0}, \qquad [6.49]$$

in particular, we can see that:

$$\sum_{i=2}^{i=p} \lambda_i \mathbf{x}_{\tau_1(i)} - \left(\sum_{i=2}^{i=p} \lambda_i b\left(\mathbf{x}_{\tau_1(i)}, \mathbf{e}_1\right)\right) q(\mathbf{e}_1)\, \mathbf{e}_1 = \mathbf{0}, \qquad [6.50]$$

that is,

$$\sum_{i=2}^{i=p} \lambda_i \mathbf{x}_{\tau_1(i)} - \left(\sum_{i=2}^{i=p} \lambda_i b\left(\mathbf{x}_{\tau_1(i)}, \mathbf{e}_1\right)\right) q(\mathbf{e}_1) \frac{\mathbf{x}_{\tau_1(1)}}{\sqrt{\left|q\left(\mathbf{x}_{\tau_1(1)}\right)\right|}} = \mathbf{0}, \qquad [6.51]$$

since the family $\mathbf{x}_1, \cdots, \mathbf{x}_p$ is linearly independent, the family $\mathbf{x}_{\tau_1(1)}, \cdots, \mathbf{x}_{\tau_1(p)}$ is also linearly independent, which entails that all the $\lambda_i, i \in [2,p]$ are null. Therefore, the family $\mathbf{x}_{i,1}, i \in [2,n]$ is a basis of $[vec(\mathbf{e}_1)]^{\perp}$, which is nondegenerate. Furthermore, we clearly have $vec(\mathbf{e}_1) = vec\left(\mathbf{x}_{\tau(1)}\right)$. Let $i \in [2,n]$ be an index such that:

$$|q(\mathbf{x}_{i,1})| = \max_{j \in [2,n]} |q(\mathbf{x}_{j,1})|, \qquad [6.52]$$

and consider a transposition τ_2 that permutes 2 and j with $j \geq 2$. Obviously,

$$q\left(\mathbf{x}_{\tau_2(2),1}\right) \neq 0, \qquad [6.53]$$

otherwise the orthogonal of $vec(\mathbf{e}_1)$ within $\boldsymbol{F}$ would be degenerate. We then set:

$$\mathbf{e}_2 = \frac{\mathbf{x}_{\tau_2(2),1}}{\sqrt{\left|q\left(\mathbf{x}_{\tau_2(2),1}\right)\right|}}. \qquad [6.54]$$

Consider now the vectors:

$$\mathbf{x}_{\tau_2(2),1}, \cdots, \mathbf{x}_{\tau_2(n),1}, \qquad [6.55]$$

$$\forall i \in [3, n], \;\; \mathbf{x}_{i,2} := \mathbf{x}_{\tau_2(i),1} - q(\mathbf{e}_1)\, b\left(\mathbf{x}_{\tau_2(i),1}, \mathbf{e}_1\right)\mathbf{e}_1 - q(\mathbf{e}_2)\, b\left(\mathbf{x}_{\tau_2(i),1}, \mathbf{e}_2\right)\mathbf{e}_2. \qquad [6.56]$$

In virtue of the same considerations as above, the family of $\mathbf{x}_{i,2}$ is a linearly independent family of $[vec(\mathbf{e}_1, \mathbf{e}_2)]^{\perp}$. Since $\mathbf{e}_1, \mathbf{e}_2$ is an orthonormal basis of $vec(\mathbf{e}_1, \mathbf{e}_2)$, this space is nondegenerate and thus its orthogonal (in $\boldsymbol{F}$) is also nondegenerate. On the other hand, we have:

$$vec(\mathbf{e}_1, \mathbf{e}_2) = vec\left(\mathbf{x}_{\tau_1(1)}, \mathbf{x}_{\tau_2(2),1}\right),$$
$$\mathbf{x}_{\tau_2(2),1} = \mathbf{x}_{\tau_1(2)} - b\left(\mathbf{x}_{\tau_1(2)}, \mathbf{e}_1\right) q(\mathbf{e}_1)\,\mathbf{e}_1. \qquad [6.57]$$

In that case, we can observe that we indeed have:

$$vec(\mathbf{e}_1, \mathbf{e}_2) = vec\left(\mathbf{x}_{\tau_2\tau_1(1)}, \mathbf{x}_{\tau_2\tau_1(2)}\right). \qquad [6.58]$$

Then, it is possible to find an anisotropic element by selecting the maximum of $|q(\mathbf{x}_{i,2})|$ for $i \in [3, n]$, and so forth up to the order p. The final permutation σ is obtained as a composition of different permutations $\sigma = \tau_{p-1} \circ \cdots \circ \tau_1$, knowing that each transposition τ_k is defined as $\tau_k(k) = q$ with $q \geq k$.

It is equally possible to extract an orthonormal basis starting from the pseudo-dot product matrix $\mathbb{B}$:

THEOREM 6.6.– Let $\boldsymbol{E}$ be a linear space of finite dimension and b a symmetric bilinear form over $\boldsymbol{E}$. Then, there exists $\mathbf{e}_1, \cdots, \mathbf{e}_n$ a basis of $\boldsymbol{E}$ and some reals $\lambda_1, \cdots, \lambda_n$ such that:

$$1)\ \forall i \neq j,\ \ b(\mathbf{e}_i, \mathbf{e}_j) = 0\ 2)\ \forall i \in [1, n],\ \ q(\mathbf{e}_i) = \lambda_i.$$

In particular, if b is nondegenerate all the λ_i are non-null. The family defined by

$$\forall i \in [1, n],\ \ \frac{1}{\sqrt{|\lambda_i|}}\mathbf{e}_i \tag{6.59}$$

is then a pseudo-orthonormal basis of $(\boldsymbol{E}, b)$.

PROOF.– Consider a space $\boldsymbol{E}$ of dimension n. Let us choose a basis $\mathbf{f}_1, \cdots, \mathbf{f}_n$ of this space. Then, we construct the real matrix:

$$\forall (i, j) \in [1, n]^2,\ \ \mathbb{B}_{ij} := b(\mathbf{f}_i, \mathbf{f}_j). \tag{6.60}$$

In that case, such a matrix is a real symmetric matrix: thus, it admits a basis (of column vectors) $E_1, \cdots, E_n$ and a set of eigenvalues $\lambda_1, \cdots, \lambda_n$ such that:

$$\forall k \in [1, n],\ \ \mathbb{B}E_k = \lambda_k E_k,\ \ \forall k, l\ \sum_{i=1}^{i=n} E_{k,i} E_{l,i} = \delta_{kl}. \tag{6.61}$$

We define then for each element k a vector $\mathbf{e}_k$ according to the relation:

$$\mathbf{e}_k = \sum_{i=1}^{i=1} E_{k,i}\mathbf{f}_i. \tag{6.62}$$

Now, let us calculate:

$$b(\mathbf{e}_k, \mathbf{e}_l) = \sum_{i,j} E_{k,i} E_{l,j} b(\mathbf{f}_i, \mathbf{f}_j) = \sum_{i,j} E_{k,i} E_{l,j} \mathbb{B}_{ij}, \tag{6.63}$$

the right-hand side is calculated as:

$$\sum_{i=1}^{i=n} E_{k,i} \left(\sum_{j=1}^{j=n} \mathbb{B}_{ij} E_{l,j} \right) = \sum_{i=1}^{i=n} E_{k,i} \lambda_l E_{l,i} \tag{6.64}$$

with the latter equality following from the fact that the vector E_l is an eigenvector of the matrix $\mathbb{B}$ for the eigenvalue λ_l. Hence, we deduce that:

$$b(\mathbf{e}_k, \mathbf{e}_l) = \lambda_l \delta_{kl}, \tag{6.65}$$

which proves our result. Now, if b is nondegenerate, the matrix $\mathbb{B}$ is invertible. None of λ_i can be null and the basis $\mathbf{e}_i/\sqrt{|\lambda_i|}$ is a pseudo-orthonormal basis of b.

We provide a corollary that is useful for the above demonstration:

COROLLARY 6.5.– Let $(\boldsymbol{E}, b)$ be a pseudo-Euclidean space. Then, any subspace $\boldsymbol{F}$ of $\boldsymbol{E}$ admits a pseudo-orthogonal basis of b.

PROOF.– The form b restricted to $\boldsymbol{F}$, denoted by $b_{\boldsymbol{F}}$, is always symmetric and thus, in virtue of the previous demonstration, we always have the existence of a pseudo-orthogonal basis of $(\mathbf{F}, b_{\mathbf{F}})$. This family is also pseudo-orthogonal for the form b.

6.2.3. *Signature of a nondegenerate quadratic form*

We now enunciate Sylvester's law, which enables the research of the definition of a signature associated with a pseudo-orthonormal basis:

THEOREM 6.7 (Sylvester's law of inertia).– Let $(\boldsymbol{E}, b)$ be a pseudo-Euclidean space. Then, for any pseudo-orthonormal basis $\mathbf{e}_i, i \in [1, n]$, the number p of vectors such that $q(\mathbf{e}_i) = 1$ and the number s of vectors such that $q(\mathbf{e}_i) = -1$ are independent of the chosen basis. The number s is called signature of the pseudo-dot product b.

PROOF.– Consider $\mathbf{e}_1, \cdots, \mathbf{e}_n$. We denote by p the number of vectors such that $q(\mathbf{e}_i) = 1$ and by s the number of vectors such that $q(\mathbf{e}_i) = -1$. Since the basis is pseudo-orthonormal, we have $p + s = n$. Consider another orthonormal basis $\mathbf{f}_1, \cdots, \mathbf{f}_n$. We denote by p', s' the number of vectors of quadratic form equal to $+1, -1$. Again, $p' + s' = n$. Even if the order is changed, we can suppose that the first vectors are of positive quadratic form and the following ones are of negative quadratic form. Up to a notation's inversion, suppose that $p > p'$. We set $\boldsymbol{F} = vec(\mathbf{e}_1, \cdots, \mathbf{e}_p)$ and $\boldsymbol{F}' = vec(\mathbf{f}_{p'+1}, \cdots, \mathbf{f}_n)$. We then have:

$$\dim(\boldsymbol{F}') + \dim(\boldsymbol{F}) = n - p' + p = n + (p - p') > n, \qquad [6.66]$$

more specifically, we necessarily have $\dim(\boldsymbol{F} \bigcap \boldsymbol{F}') > 0$. In fact, we always have:

$$\dim(\boldsymbol{F} + \boldsymbol{F}') \leq n \text{ and } \dim(\boldsymbol{F} + \boldsymbol{F}') = \dim(\boldsymbol{F}') + \dim(\boldsymbol{F}) - \dim(\boldsymbol{F} \bigcap \boldsymbol{F}'), \qquad [6.67]$$

which proves that we have:

$$\dim(\boldsymbol{F} \bigcap \boldsymbol{F}') \geq (n + (p - p')) - n \geq p - p' > 0, \qquad [6.68]$$

therefore there exists a linear subspace not reduced to $\{\mathbf{0}\}$ such that b is strictly positive and strictly negative over this space: absurd.

Let us conclude with a technical albeit useful theorem:

THEOREM 6.8.– Let $\boldsymbol{E}, b$ be a space with a dimension n and a signature s. Then, the following statements are true:

1) a larger linear subspace $\boldsymbol{F}^+$ such that the restriction of q to $\boldsymbol{F}^+$ is (strictly) positive over $\boldsymbol{F}^+$ has a dimension $n - s$;

2) a larger linear subspace $\boldsymbol{F}^-$ such that the restriction of q to $\boldsymbol{F}^-$ is (strictly) negative over $\boldsymbol{F}-$ has a dimension s;

3) a larger linear subspace $\boldsymbol{F}^0$ such that the restriction of q to $\boldsymbol{F}^0$ is (exactly) null has a dimension $\min(s, n - s)$.

PROOF.– Let us demonstrate these properties:

1) We show the first two points simultaneously. Let $\mathbf{e}_1, \cdots, \mathbf{e}_n$ be a pseudo-orthonormal basis of $(\mathbf{E}, \mathbf{b})$. Then, according to Sylvester's law of inertia, it is possible to extract $\mathbf{e}_1, \cdots, \mathbf{e}_s$ such that $q(\mathbf{e}_i) = -1,\ i \in [1, s]$ and $\mathbf{e}_{s+1}, \cdots, \mathbf{e}_n$ such that $q(\mathbf{e}_j) = 1,\ j \in [s+1, n]$. By considering then the spaces $\boldsymbol{F}^- = \langle \mathbf{e}_1, \cdots, \mathbf{e}_s \rangle$ and $\boldsymbol{F}^+ = \langle \mathbf{e}_{s+1}, \cdots, \mathbf{e}_n \rangle$, we can observe the existence of desired subspaces with dimensions s and $n - s$. Suppose now that there exists a subspace $\boldsymbol{F}^-$ with a dimension strictly greater that s. Clearly, $\boldsymbol{F}^-$ is nondegenerate. We can find a pseudo-orthonormal basis of $\boldsymbol{F}^-$, which can be completed in order to obtain a pseudo-orthonormal basis of $\boldsymbol{E}$. Such a pseudo-orthonormal basis has then at least $s + 1$ vectors which are strictly negative for b: therefore, it contradicts Sylvester's law of inertia. Evidently, the demonstration for the spaces $\boldsymbol{F}^+$ is completely identical.

2) Let $\boldsymbol{F}^+$ be a linear subspace of dimension $n - s$. Clearly, it admits a unique orthogonal supplementary subspace of type $\boldsymbol{F}^-$ with a dimension s. Let $l = \min(s, n - s)$: by choosing some vectors of $\boldsymbol{F}^+, \boldsymbol{F}^-$, it is possible to construct a family:

$$\mathbf{e}_1^+, \mathbf{e}_1^-, \cdots, \mathbf{e}_l^+, \mathbf{e}_l^-, \text{ such that } q(\mathbf{e}_i^+) = -q(\mathbf{e}_i^-) = 1, b(\mathbf{e}_i^+, \mathbf{e}_i^-) = 0. \quad [6.69]$$

In that case, the family $\mathbf{e}_i := \mathbf{e}_i^+ + \mathbf{e}_i^-,\ i \in [1, l]$ is a linearly independent family of $\boldsymbol{E}$ that furthermore satisfies $\forall i \in [1, l],\ q(\mathbf{e}_i) = 0$. The existence of an isotropic subspace with a dimension $\min(s, n - s)$ holds. If an isotropic subspace with a greater dimension existed, due to dimension limitations, it would necessarily intersect a subspace of type $\boldsymbol{F}^+$ or $\boldsymbol{F}^-$ with a dimension s or $n - s$: this is, however, impossible.

7

Pseudo-Euclidean Algebras

In this chapter, we answer the following question:

QUESTION 7.1.– Let $(\boldsymbol{E}, b)$ be a pseudo-Euclidean space. Is it possible to construct a pseudo-dot product over $\Lambda^p \boldsymbol{E}$ in a natural way?

The answer to this question is quite interesting, as it combines the notion of determinant with the application of pseudo-dot product.

7.1. Pseudo-dot product over $\Lambda^p E$

7.1.1. *Construction*

THEOREM 7.1.– Let $(\boldsymbol{E}, b)$ be a pseudo-Euclidean space and $p \geq 1$. Then, there exists a unique pseudo-dot product $B_p(\cdot,\cdot)$ over $\Lambda^p \boldsymbol{E}$ such that:

$$\forall (\mathbf{x}_1, \cdots, \mathbf{x}_p) \in \boldsymbol{E}^p, \quad \forall (\mathbf{y}_1, \cdots, \mathbf{y}_p) \in \boldsymbol{E}^p,$$

$$B_p(\mathbf{x}_1 \wedge \cdots \wedge \mathbf{x}_p, \mathbf{y}_1 \wedge \cdots \wedge \mathbf{y}_p) = \det(b(\mathbf{x}_i, \mathbf{y}_j)). \qquad [7.1]$$

REMARK 7.1.– The case $p = 1$ is trivial since $B_1 = b$, and it is possible to "extend" the definition to the case $p = 0$ by setting B_0 to represent a multiplication between real numbers. ■

DEFINITION 7.1.– Let $(\boldsymbol{E}, b)$ be a pseudo-Euclidean space. We call pseudo-dot product induced over $\Lambda^p \boldsymbol{E}$ the form B_p from the previous theorem.

Let us now address the demonstration of the theorem:

PROOF.– Let $\mathbf{x}_1, \cdots, \mathbf{x}_p$ be p arbitrarily fixed vectors of $\boldsymbol{E}$. Then, the map defined as

$$(\mathbf{y}_1, \cdots, \mathbf{y}_p) \mapsto \det\left(b\left(\mathbf{x}_i, \mathbf{y}_j\right)\right) \qquad [7.2]$$

is clearly a skew-symmetric form over $\boldsymbol{E}^p$, because the map $b(\mathbf{z}, \cdot)$ is linear for a fixed $\mathbf{z}$, and the map det is an alternating p-linear map. According to the universal theorem of construction of the exterior algebra (see theorem 2.1), and by employing the symbol "$\wedge$" as a construction operator, there exists a unique linear form $l[\mathbf{x}_1, \cdots, \mathbf{x}_p](\cdot)$ over $\Lambda^p \boldsymbol{E}$ such that:

$$\forall (\mathbf{y}_1, \cdots, \mathbf{y}_p) \in \boldsymbol{E}^p, \;\; l[\mathbf{x}_1, \cdots, \mathbf{x}_p](\mathbf{y}_1 \wedge \cdots \wedge \mathbf{y}_p) = \det\left(b\left(\mathbf{x}_i, \mathbf{y}_j\right)\right). \qquad [7.3]$$

Consider now the map $\beta : \boldsymbol{E}^p \mapsto \mathcal{L}(\Lambda^p \boldsymbol{E}, \mathbb{R})$ defined as:

$$\forall (\mathbf{x}_1, \cdots, \mathbf{x}_p) \in \boldsymbol{E}^p, \;\; \beta(\mathbf{x}_1, \cdots, \mathbf{x}_p) = l[\mathbf{x}_1, \cdots, \mathbf{x}_p](\cdot). \qquad [7.4]$$

Then, for all $\sigma \in S_p$ and for all $\mathbf{y}_1 \wedge \cdots \wedge \mathbf{y}_p$, we have:

$$l\left[\mathbf{x}_{\sigma(1)}, \cdots, \mathbf{x}_{\sigma(p)}\right](\mathbf{y}_1 \wedge \cdots \wedge \mathbf{y}_p) = \det\left(b\left(\mathbf{x}_{\sigma(i)}, \mathbf{y}_j\right)\right) = \varepsilon(\sigma) \det\left(b\left(\mathbf{x}_i, \mathbf{y}_j\right)\right). \qquad [7.5]$$

Hence, we deduce that:

$$l\left[\mathbf{x}_{\sigma(1)}, \cdots, \mathbf{x}_{\sigma(p)}\right](\mathbf{y}_1 \wedge \cdots \wedge \mathbf{y}_p) = \varepsilon(\sigma)\, l[\mathbf{x}_1, \cdots, \mathbf{x}_p](\mathbf{y}_1 \wedge \cdots \wedge \mathbf{y}_p), \qquad [7.6]$$

since the equality above is true for all the exterior products of the form $\mathbf{y}_1 \wedge \cdots \wedge \mathbf{y}_p$, it holds as well for all the vectors $\mathbf{Y}$ of $\Lambda^p \boldsymbol{E}$ (according to proposition 4.1 and in virtue of the linearity of the map $l[\mathbf{x}_1, \cdots, \mathbf{x}_p]$ over $\Lambda^p \boldsymbol{E}$). Therefore, we deduce that:

$$\forall \sigma \in S_p, \;\; \beta\left(\mathbf{x}_{\sigma(1)}, \cdots, \mathbf{x}_{\sigma(p)}\right) = \varepsilon(\sigma)\, \beta(\mathbf{x}_1, \cdots, \mathbf{x}_p). \qquad [7.7]$$

Furthermore, it can easily be seen that the map β is multi-linear (due to the multi-linearity of the determinant and to the bilinearity of the form b). As a result, according to the fundamental theorem of construction of the exterior algebra (theorem 2.1), there exists a unique linear application $\mathcal{B}_p : \Lambda^p \boldsymbol{E} \mapsto \mathcal{L}(\Lambda^p \boldsymbol{E}, \mathbb{R})$ such that:

$$\forall (\mathbf{x}_1, \cdots, \mathbf{x}_p) \in \boldsymbol{E}^p, \;\; \mathcal{B}_p(\mathbf{x}_1 \wedge \cdots \wedge \mathbf{x}_p)(\cdot) = l[\mathbf{x}_1, \cdots, \mathbf{x}_p](\cdot). \qquad [7.8]$$

Moreover, it is evident, thanks to proposition 1.13, that the space of bilinear forms over a space $\boldsymbol{X}$ is isomorphic to the space $\mathcal{L}(\boldsymbol{X}, \mathcal{L}(\boldsymbol{X}, \mathbb{R}))$: a linear map from $\boldsymbol{X}$ with values in linear forms over $\boldsymbol{X}$ defines a bilinear form univocally (and thus the reciprocal is true as well). Therefore, there exists a unique bilinear form $B_p(\cdot,\cdot)$ over $\Lambda^p\boldsymbol{E}$ such that:

$$B_p(\mathbf{x}_1 \wedge \cdots \wedge \mathbf{x}_p, \mathbf{y}_1 \wedge \cdots \wedge \mathbf{y}_p) = \det(b(\mathbf{x}_i, \mathbf{y}_j)), \quad [7.9]$$

clearly, such a map is symmetric: inverting $\mathbf{x}_i, \mathbf{y}_j$ is equivalent to calculating the determinant of the transpose matrix rather than that of the matrix itself, since they are equal (see proposition 5.2 regarding the determinant of transpose matrices). Moreover, such a map is nondegenerate. In fact, let us choose a pseudo-orthonormal basis $\mathbf{e}_1, \cdots, \mathbf{e}_n$ of the space $\boldsymbol{E}$ and let us construct the basis:

$$\mathbf{E}_\mathbf{i} := \mathbf{e}_{i_1} \wedge \cdots \wedge \mathbf{e}_{i_p}, \quad \mathbf{i} := i_1, \cdots i_p \in \mathcal{I}_o \quad [7.10]$$

Then, we have, for every pair $\mathbf{i}, \mathbf{j} \in \mathcal{I}_o$:

$$B_p(\mathbf{E}_\mathbf{i}, \mathbf{E}_\mathbf{j}) = \det(b(\mathbf{e}_{i_k}, \mathbf{e}_{j_l})), \quad k, l \in [1, p]^2. \quad [7.11]$$

We can observe the following points:

1) Suppose that $i_1 \neq j_1$. If necessary, we can also suppose that $i_1 < j_1$. Then, since the multi-index $\mathbf{j}$ is strictly ordered, we have $\forall l \in [1, p], i_1 < j_1 \leq j_p$. Hence, we deduce that $\forall l \in [1, p],\ i_1 \neq j_p$ which implies that $\forall l \in [1, p], b(\mathbf{e}_{i_1}, \mathbf{e}_{j_l}) = 0$ since the basis $\mathbf{e}_1, \cdots, \mathbf{e}_n$ is pseudo-orthonormal. Therefore, the first row of the matrix $b(\mathbf{e}_{i_k}, \mathbf{e}_{j_l})$ is null and its determinant is null as well. Thus, $i_1 \neq j_1$ entails that $B_p(\mathbf{E}_\mathbf{i}, \mathbf{E}_\mathbf{j}) = 0$.

2) Suppose now that $i_1 = j_1$. Then, the first row (and, by symmetry, first column) of the matrix $b(\mathbf{e}_{i_k}, \mathbf{e}_{j_l})$ is such that the first term equals $b(\mathbf{e}_{i_1}, \mathbf{e}_{i_1})$ and the other terms equal 0. We can thus write that:

$$\det(b(\mathbf{e}_{i_k}, \mathbf{e}_{j_l})), \quad k, l \in [1, p]^2 = q(\mathbf{e}_{i_1}) \det(b(\mathbf{e}_{i_k}, \mathbf{e}_{j_l})), \quad k, l \in [2, p]^2. \quad [7.12]$$

3) If $i_2 \neq j_2$, we apply the same arguments as above and we obtain: $B_p(\mathbf{E}_\mathbf{i}, \mathbf{E}_\mathbf{j}) = 0$.

4) Otherwise, for $i_2 = j_2$, we obtain:

$$\det(b(\mathbf{e}_{i_k}, \mathbf{e}_{j_l})), \quad k, l \in [1, p]^2 = q(\mathbf{e}_{i_1})\, q(\mathbf{e}_{i_2}) \det(b(\mathbf{e}_{i_k}, \mathbf{e}_{j_l})), \quad k, l \in [3, p]^2. \quad [7.13]$$

5) By induction, we obtain the following result:

$$\forall \mathbf{i}, \mathbf{j} \in \mathcal{I}_o, \quad B_p(\mathbf{E}_\mathbf{i}, \mathbf{E}_\mathbf{j}) = \begin{cases} 0 \ \mathit{if}\ \mathbf{i} \neq \mathbf{j} \\ q(\mathbf{e}_{i_1}) \cdots q(\mathbf{e}_{i_p}) \ \ \mathit{if}\ \mathbf{i} = \mathbf{j} \end{cases}. \quad [7.14]$$

Since the basis $\mathbf{e}_1, \cdots, \mathbf{e}_n$ is pseudo-orthonormal for the form b, this implies that all the $q(\mathbf{e}_i)$ are non-null (in fact, they are all equal to ± 1). In particular, it can immediately be seen that the matrix of the form B_p in the basis of $\mathbf{E}_{\mathrm{i}}$ is a diagonal matrix (and thus symmetric) with non-null terms along its diagonal. Therefore, this matrix is indeed invertible, which proves, in accordance with proposition 6.4, that the form B_p is indeed nondegenerate over $\Lambda^p \boldsymbol{E}$.

7.1.2. *Calculations and applications*

A number of statements stem from the previous calculations:

COROLLARY 7.1.– Given any pseudo-orthonormal basis $\mathbf{e}_1, \cdots, \mathbf{e}_n$ of $(\boldsymbol{E}, b)$, the basis:

$$\mathbf{E}_{\mathrm{i}} := \mathbf{e}_{i_1} \wedge \cdots \wedge \mathbf{e}_{i_p}, \quad 1 \leq i_1 < \cdots < i_p \leq n \tag{7.15}$$

is a pseudo-orthonormal basis of $(\Lambda^p \boldsymbol{E}, B_p)$ and, moreover, we have:

$$\forall \mathbf{X} = \sum_{\mathrm{i} \in \mathcal{I}_o} X_{\mathrm{i}} \mathbf{E}_{\mathrm{i}}, \quad \forall \mathbf{Y} = \sum_{\mathrm{i} \in \mathcal{I}_o} Y_{\mathrm{i}} \mathbf{E}_{\mathrm{i}}, \quad B_p(\mathbf{X}, \mathbf{Y}) = \sum_{\mathrm{i} \in \mathcal{I}_o} X_{\mathrm{i}} Y_{\mathrm{i}} Q_p(\mathbf{E}_{\mathrm{i}}), \tag{7.16}$$

where, explicitly,

$$Q_p(\mathbf{E}_{\mathrm{i}}) = q(\mathbf{e}_{i_1}) \cdots q(\mathbf{e}_{i_p}). \tag{7.17}$$

PROOF.– The demonstration of the previous theorem showed that the basis of the $\mathbf{E}_{\mathrm{i}}$ was actually pseudo-orthogonal. The calculation of the $Q(\mathbf{E}_{\mathrm{i}})$ completed the demonstration that this basis was in fact pseudo-orthonormal.

A practical method of calculating the quadratic form of a vector obtained as an exterior product of vectors follows:

PROPOSITION 7.1.– Let $\mathbf{X} = \mathbf{x}_1 \wedge \cdots \wedge \mathbf{x}_p$ and let $\mathbf{e}_1, \cdots, \mathbf{e}_n$ be a basis $(\boldsymbol{E}, b)$. We denote by X the matrix (with n rows and p columns) of the components of $\mathbf{x}_1, \cdots, \mathbf{x}_p$ within the basis $\mathbf{e}_1, \cdots, \mathbf{e}_n$. Furthermore, we denote by Γ the matrix of $b(\mathbf{e}_i, \mathbf{e}_j)$. Then, we have:

$$Q_p(\mathbf{X}) = Q_p(\mathbf{x}_1 \wedge \cdots \wedge \mathbf{x}_p) = \det\left(X^{\mathsf{T}} \Gamma X\right). \tag{7.18}$$

PROOF.– It consists of a simple writing exercise by employing the previous corollary:

$$Q_p(\mathbf{x}_1 \wedge \cdots \wedge \mathbf{x}_p) = \det(b(\mathbf{x}_i, \mathbf{x}_j)), \quad (i, j) \in [1, p]^2. \tag{7.19}$$

We develop the expression by writing that:

$$\mathbf{x}_i := \sum_{k=1}^{k=n} X_{k,i}\mathbf{e}_k, \qquad [7.20]$$

hence, we deduce that:

$$Q_p(\mathbf{x}_1 \wedge \cdots \wedge \mathbf{x}_p) = \det\left(\sum_{k=1}^{k=n} X_{k,i} b(\mathbf{e}_k, \mathbf{e}_l) \sum_{l=1}^{l=n} X_{l,j}\right). \qquad [7.21]$$

In matrix form, this is actually equivalent to writing that:

$$Q_p(\mathbf{x}_1 \wedge \cdots \wedge \mathbf{x}_p) = \det\left(X^\top \Gamma X\right). \qquad [7.22]$$

A characterization of degenerate subspaces is a consequence of the previous proposition:

COROLLARY 7.2.– Let $\boldsymbol{F}$ be a subspace of $(\boldsymbol{E}, b)$ with a dimension p. Then, $\boldsymbol{F}$ is degenerate if and only if for all family $\mathbf{x}_1, \cdots, \mathbf{x}_p$ of $\boldsymbol{F}$, we have:

$$Q_p(\mathbf{x}_1 \wedge \cdots \wedge \mathbf{x}_p) = 0.$$

PROOF.– If the family $\mathbf{x}_1, \cdots, \mathbf{x}_p$ is linearly dependent, then $\mathbf{x}_1 \wedge \cdots \wedge \mathbf{x}_p = \mathbf{0}$ and thus necessarily $Q_p(\mathbf{x}_1 \wedge \cdots \wedge \mathbf{x}_p) = 0$. Otherwise, the family $\mathbf{x}_1, \cdots, \mathbf{x}_p$ is a basis of $\boldsymbol{F}$. We complete it by a family $\mathbf{x}_{p+1}, \cdots, \mathbf{x}_n$ to obtain a basis of $\boldsymbol{E}$. Let Γ be the matrix of b with respect to the basis $\mathbf{x}_1, \cdots, \mathbf{x}_n$ and let $\Gamma_{\boldsymbol{F}}$ be the square matrix with a size p obtained from $b(\mathbf{x}_i, \mathbf{x}_j), i, j \in [1, p]^2$. In other words, we set:

$$\forall i, j \in [1, n]^2, [\Gamma]_{i,j} = b(\mathbf{x}_i, \mathbf{x}_j), \quad \forall i, j \in [1, p], [\Gamma_{\boldsymbol{F}}]_{i,j} = b(\mathbf{x}_i, \mathbf{x}_j). \qquad [7.23]$$

According to the previous proposition, we have:

$$Q_p(\mathbf{x}_1 \wedge \cdots \wedge \mathbf{x}_p) = \det\left(X^\top \Gamma X\right) = \det(\Gamma_{\boldsymbol{F}}), \qquad [7.24]$$

due to the fact that:

$$X = \begin{bmatrix} \mathbb{I}_p \\ \mathbb{O}_{n-p,p} \end{bmatrix} \Rightarrow X^\top \Gamma X = \Gamma_{\boldsymbol{F}}. \qquad [7.25]$$

Now, according to proposition 6.4, the space $\boldsymbol{F}$ is degenerate if and only if the matrix $\Gamma_{\boldsymbol{F}}$ admits at least one null eigenvalue, thus if and only if its determinant is null (actually, since the matrix Γ_F is a symmetric real matrix, it is diagonalizable and its determinant is equal to the product of its eigenvalues).

COROLLARY 7.3.– Let $\mathbf{x}_1, \cdots, \mathbf{x}_p$ be a family such that:

$$Q_p(\mathbf{x}_1 \wedge \cdots \wedge \mathbf{x}_p) \neq 0,$$

then, the family $\mathbf{x}_1, \cdots, \mathbf{x}_p$ is linearly independent.

PROOF.– Let $\mathbf{x}_1, \cdots, \mathbf{x}_p$ be a family of $\boldsymbol{F}$. Then, we know that:

$$Q_p(\mathbf{x}_1 \wedge \cdots \wedge \mathbf{x}_p) = \det \mathbb{B}, \quad \mathbb{B}_{ij} = b(\mathbf{x}_i, \mathbf{x}_j). \tag{7.26}$$

Suppose now that:

$$\sum_{j=1}^{j=n} \lambda_j \mathbf{x}_j = 0, \tag{7.27}$$

then, we deduce thereof that:

$$\forall i \in [1, n], \quad \sum_{j=1}^{j=n} b(\mathbf{x}_i, \mathbf{x}_j) \lambda_j = 0, \tag{7.28}$$

however, the hypothesis of the corollary is precisely that the matrix $\mathbb{B}$ has a non-null determinant and thus is invertible. As a result, the only solution to the linear system above is that all the λ_j are null.

REMARK 7.2.– In the pseudo-Euclidean case, the reciprocal statement is, in general, false: it suffices to choose $\mathbf{x} \neq 0$ such that $q(\mathbf{x}) = 0$. On the other hand, in the Euclidean case the reciprocal assertion is always true and even trivial: if $Q_p^+(\mathbf{x}_1 \wedge \cdots \wedge \mathbf{x}_p) \neq 0$, then this entails that $\mathbf{x}_1 \wedge \cdots \wedge \mathbf{x}_p \neq \mathbf{0}$ and thus that the family is linearly independent. In the pseudo-Euclidean case, the reciprocal statement is weakened in the following way: ■

COROLLARY 7.4.– Let $\boldsymbol{F}$ be a subspace of $(\boldsymbol{E}, b)$ with a dimension p. Then, $\boldsymbol{F}$ is nondegenerate if and only if there exists a family $\mathbf{x}_1, \cdots, \mathbf{x}_p$ such that we have $Q_p(\mathbf{x}_1 \wedge \cdots \wedge \mathbf{x}_p) \neq 0$. In such case, for all linearly independent family $\mathbf{x}_1, \cdots, \mathbf{x}_p$ of $\boldsymbol{F}$, we have $Q_p(\mathbf{x}_1 \wedge \cdots \wedge \mathbf{x}_p) \neq 0$.

PROOF.– The first part of the statement corresponds exactly to the previous corollary. As for the second part, if $\boldsymbol{F}$ is nondegenerate, then it is possible to find a pseudo-orthonormal basis $\mathbf{e}_1, \cdots, \mathbf{e}_p$ of $(\boldsymbol{F}, b_{\boldsymbol{F}})$. In that case, for all linearly independent family $\mathbf{x}_1, \cdots, \mathbf{x}_p$ of $\boldsymbol{F}$, we have:

$$\mathbf{x}_1 \wedge \cdots \wedge \mathbf{x}_p = \det_{\mathbf{e}_1, \cdots, \mathbf{e}_p} (\mathbf{x}_1, \cdots, \mathbf{x}_p) \, \mathbf{e}_1 \wedge \cdots \wedge \mathbf{e}_p, \quad \det_{\mathbf{e}_1, \cdots, \mathbf{e}_p} (\mathbf{x}_1, \cdots, \mathbf{x}_p) \neq 0. \qquad [7.29]$$

Hence, we deduce:

$$\begin{aligned} Q_p (\mathbf{x}_1 \wedge \cdots \wedge \mathbf{x}_p) &= \left[\det_{\mathbf{e}_1, \cdots, \mathbf{e}_p} (\mathbf{x}_1, \cdots, \mathbf{x}_p) \right]^2 Q_p (\mathbf{e}_1 \wedge \cdots \wedge \mathbf{e}_p) \\ &= \left[\det_{\mathbf{e}_1, \cdots, \mathbf{e}_p} (\mathbf{x}_1, \cdots, \mathbf{x}_p) \right]^2 q (\mathbf{e}_1) \cdots q (\mathbf{e}_p). \end{aligned} \qquad [7.30]$$

Each term of the last product is non-null, which concludes the demonstration.

The theorem of characterization of linearly independent families can thus be enunciated:

THEOREM 7.2.– Let $\boldsymbol{F}$ be a nondegenerate subspace with a dimension p. Then, we have:

$\mathbf{x}_1, \cdots, \mathbf{x}_p$ *is a linearly independent family of* $\boldsymbol{F} \Leftrightarrow Q_p (\mathbf{x}_1 \wedge \cdots \wedge \mathbf{x}_p) \neq 0.$ [7.31]

7.2. Applications: volume and infinitesimal volumes

In this section, the discourse is restricted *to Euclidean spaces*. Among other features, the dot product defines a metrics that enables in its turn a natural definition of topology. The Borel-Lebesgue measure (also known as Heine-Borel measure) is obtained by considering the σ-algebra of the open sets, and the measure of the Cartesian products of closed intervals as the product of their lengths. We do not mention the issue, usually seldom addressed, of knowing which kind of topology admits linear spaces equipped with a pseudo-dot product and which measures can be associated with them.

7.2.1. *Measure of a parallelotope*

DEFINITION 7.2.– Let $\mathbf{x}_1, \cdots, \mathbf{x}_p$ be a linearly independent family of p vectors of a space $\boldsymbol{E}$ with a dimension $n \geq p \geq 1$. We call parallelotope constructed over the family $\mathbf{x}_1, \cdots, \mathbf{x}_p$ the set:

$$\mathcal{P}(\mathbf{x}_1, \cdots, \mathbf{x}_p) = \left\{ \sum_{i=1}^{i=p} u_i \mathbf{x}_i, u_i \in [0,1] \right\}. \qquad [7.32]$$

In dimension one, the parallelotope is a segment, in dimension two, it is a parallelogram, in dimension three a parallelepiped, etc. An important issue in geometry concerns the calculation of a measure (length, area and volume) of such a geometric set. The following theorem provides an answer to such a problem:

THEOREM 7.3.– Let $\mathcal{P}(\mathbf{x}_1, \cdots, \mathbf{x}_p)$ be a linearly independent family of p vectors included in a Euclidean space $\boldsymbol{E}$ with a dimension $n \geq p$. Then, the volume of the parallelotope constructed over the vectors $\mathbf{x}_1, \cdots, \mathbf{x}_p$ is given by:

$$\mathcal{V}(\mathcal{P}(\mathbf{x}_1, \cdots, \mathbf{x}_p)) = \|\mathbf{x}_1 \wedge \cdots \wedge \mathbf{x}_p\|, \qquad [7.33]$$

where the function $\|\cdot\|$ represents the Euclidean norm derived over $\Lambda^p \boldsymbol{E}$ from the dot (not simply pseudo-dot) product inherited from the positive-definite quadratic form $Q_p(\cdot) \in \mathbb{R}^+$.

REMARK 7.3.– Before undertaking the demonstration of this theorem, some remarks should be mentioned:

1) In order to demonstrate this theorem, it is necessary, first of all, to have a measure theory for Euclidean spaces. One among the interests of defining the dot product in the way we have done it, is that the inherited metrics (i.e. the Euclidean norm) over the spaces $\Lambda^p \boldsymbol{E}$ is "compatible" with the metrics defined over the spaces $\boldsymbol{E}$. Indeed, this metrics enables the definition of measures of geometric subsets.

2) We are thus adopting a point of view of differential geometry. We remind that if a variety (with a dimension p, within $\mathbb{R}^n$) is defined by its functions of coordinates written with respect to parameters:

$$x_1(u_1, \cdots, u_p), \cdots, x_n(u_1, \cdots, u_p), (u_1 \cdots u_p) \in U, \qquad [7.34]$$

then the (Riemann) measure of the variety is given by the expression:

$$\mathcal{V} = \int_{\mathbf{u} \in U} \sqrt{\det({}^t\mathbf{M}(\mathbf{u})\mathbf{M}(\mathbf{u}))} d\mathbf{u}, \qquad [7.35]$$

with $\mathbf{M}(\mathbf{u}) := M_{ij}(\mathbf{u}), (i,j) \in [1,n] \times [1,p]$ the matrix of tangent vectors, in other words, defined as:

$$\forall i,j \in [1,n] \times [1,p], \ M_{ij}(\mathbf{u}) = \frac{\partial x_i}{\partial u_j}(\mathbf{u}). \qquad [7.36]$$

There are some conditions for this definition regarding the functions x_i and the parameters $\mathbf{u} \in U$. These conditions are all satisfied for a parallelotope on condition that the family $\mathbf{x}_1, \cdots, \mathbf{x}_p$ is linearly independent. ■

PROOF.– In a first phase, it is obvious that via Fubini's theorem, the integration yields immediately:

$$\int_{\mathbf{u} \in U} \sqrt{\det({}^t\mathbf{M}(\mathbf{u})\mathbf{M}(\mathbf{u}))} d\mathbf{u} = \int_{u_1=0}^{u_1=1} \cdots \int_{u_p=0}^{u_p=1} \sqrt{\det({}^t\mathbf{M}(\mathbf{u})\mathbf{M}(\mathbf{u}))} du_1 \cdots du_p. \qquad [7.37]$$

Being the case of a parameterization of the parallelotope, it is easily observed that we actually have:

$$\forall i \in [1,n], \ x_\varepsilon(u_1, \cdots, u_p) = \sum_{k=1}^{k=p} u_k \mathbf{x}_k \cdot \mathbf{e}_i, \ \mathbf{e}_1, \cdots, \mathbf{e}_n \ \textit{orthonormal basis} \qquad [7.38]$$

to the extent that, considering the matrix of tangent vectors, we have:

$$\forall i,j \in [1,n] \times [1,p], \ M_{ij} = \frac{\partial x_i}{\partial u_j} = \mathbf{x}_j \cdot \mathbf{e}_i \ \textit{independent of} \ \mathbf{u}, \qquad [7.39]$$

as a result, we immediately obtain:

$$\int_{u_1=0}^{u_1=1} \cdots \int_{u_p=0}^{u_p=1} \sqrt{\det({}^t\mathbf{M}\mathbf{M})} du_1 \cdots du_p = \sqrt{\det({}^t\mathbf{M}\mathbf{M})} \int_{u_1=0}^{u_1=1} \cdots \int_{u_p=0}^{u_p=1} du_1 \cdots du_p = \sqrt{\det({}^t\mathbf{M}\mathbf{M})}. \qquad [7.40]$$

Now, let us calculate the (square) matrix ${}^t\mathbf{M}\mathbf{M}$. Its element in the row $i \in [1,p]$ and the column $j \in [1,p]$ is the term:

$$\left({}^t\mathbf{M}\mathbf{M}\right)_{ij} = \sum_{k=1}^{k=n} \mathbf{x}_i \cdot \mathbf{e}_k \mathbf{x}_j \cdot \mathbf{e}_k = \mathbf{x}_i \cdot \mathbf{x}_j. \qquad [7.41]$$

According to the formula of dot product (see theorem 7.1), we know that:

$$\det\left({}^t\mathbf{M}\mathbf{M}\right) = Q_p^+\left(\mathbf{x}_1 \wedge \cdots \wedge \mathbf{x}_p\right). \qquad [7.42]$$

The theorem is thus demonstrated considering that in the Euclidean case, we have by definition $\|\mathbf{E}\| = \sqrt{Q_p^+(\mathbf{E})}$.

7.2.2. *Measure of a simplex*

Another important geometric entity defined over a linearly independent family $\mathbf{x}_1, \cdots, \mathbf{x}_p$ of p vectors of $\boldsymbol{E}$ is the simplex:

DEFINITION 7.3.– Let $\mathbf{x}_1 \cdots, \mathbf{x}_p$ be a linearly independent family of p vectors of $\boldsymbol{E}$. We call simplex constructed over this family the set $\mathcal{S}$, defined as:

$$\mathcal{S}\left(\mathbf{x}_1, \cdots, \mathbf{x}_p\right) = \left\{\sum_{i=1}^{i=p} u_i \mathbf{x}_i, \ \forall i \in [1, p], \ 0 \le u_i \le 1, \ \sum_{i=1}^{i=p} u_i = 1\right\}. \qquad [7.43]$$

The measure of a simplex can be calculated with the same approach as the one described above. Therefore, we can state that:

PROPOSITION 7.2.– Let $\mathbf{x}_1 \cdots, \mathbf{x}_p$ be a linearly independent family of p vectors of $\boldsymbol{E}$. The measure of the simplex associated with this family is expressed as:

$$\mathcal{V}\left(\mathcal{S}\left(\mathbf{x}_1, \cdots, \mathbf{x}_p\right)\right) = \frac{1}{p!} \left\|\mathbf{x}_1 \wedge \cdots \wedge \mathbf{x}_p\right\|. \qquad [7.44]$$

PROOF.– Technically, the volume calculation by means of Riemann's measure associated with the parametric representation is equivalent to the calculation performed for the parallelotope: the only difference being the limitations of parameters $u_1, \cdots, u_p$; thus, we obtain immediately:

$$\mathcal{V}\left(\mathcal{S}\left(\mathbf{x}_1, \cdots, \mathbf{x}_p\right)\right) = \left\|\mathbf{x}_1 \wedge \cdots \wedge \mathbf{x}_p\right\| \int_{u_1, \cdots u_p \in U} du_1 \cdots du_p, \qquad [7.45]$$

where U is the domain expressed as:

$$U = \left\{\left(u_1, \cdots, u_p\right), \ 0 \le u_i \le 1, \ \sum_{i=1}^{i=p} u_i = 1\right\}. \qquad [7.46]$$

We can immediately verify that this domain can be expressed piecewise by writing:

$$(u_1, \ldots, u_p) \in U \Leftrightarrow \quad \forall k \in [1, p],\; 0 \leq u_k \leq 1 - \left(\sum_{i=1}^{i=k-1} u_i \right). \qquad [7.47]$$

Therefore, we perform a sequence of integrations for each variable starting from u_p and moving back along the indexes up to u_1:

$$\int_{u_p=0}^{u_p=1-(u_1+\cdots+u_{p-2})} du_p = 1 - (u_1 + \cdots + u_{p-1}), \qquad [7.48]$$

then:

$$\int_{u_{p-1}=0}^{u_{p-1}=1-(u_1+\cdots+u_{p-2})} (1 - (u_1 + \cdots + u_{p-1}))\, du_{p-1}$$
$$= 1 - (u_1 + \cdots + u_{p-2}) - \frac{1}{2}\left(1 - (u_1 + \cdots + u_{p-2})^2\right), \qquad [7.49]$$

which can be rewritten in the form:

$$\frac{1}{2}\left(1 - (u_1 + \cdots + u_{p-2})\right)^2. \qquad [7.50]$$

We suppose that the kth integration has the form:

$$\frac{1}{k!} \int_{u_{p-k}=0}^{u_{p-k}=1-(u_1+\cdots+u_{p-k-1})} \left(1 - (u_1 + \cdots + u_{p-k})\right)^k du_{p-k}. \qquad [7.51]$$

From the primitive function of the integrand, we easily obtain:

$$\frac{1}{k!}\left(-\frac{1}{k+1}\right)\left[\left(1 - (u_1 + \cdots + u_{p-k})\right)^{k+1}\right]_{u_{p-k}=0}^{u_{p-k}=1-(u_1+\cdots+u_{p-k-1})}, \qquad [7.52]$$

which immediately yields:

$$\frac{1}{(k+1)!}\left(1 - (u_1 + \cdots + u_{p-k-1})\right)^{k+1}, \qquad [7.53]$$

hence the property of the following rank. Finally, we have indeed:

$$\int_U du_1 \cdots d_{u_p} = \frac{1}{p!}, \qquad [7.54]$$

and the result is demonstrated.

REMARK 7.4.– In a general way, we can find the volume of whichever geometric entity linearly constructed over independent vectors. If $\mathcal{G}$ is a set of the form:

$$\mathcal{G}(\mathbf{x}_1,\cdots,\mathbf{x}_p)=\sum_{i=1}^{i=p}u_i\mathbf{x}_i,\quad (u_1,\cdots,u_p)\in U \tag{7.55}$$

with $\mathbf{x}_1,\cdots,\mathbf{x}_p$ a linearly independent family, then we have, similarly to the described method:

$$\mathcal{V}(\mathcal{G}(\mathbf{x}_1,\cdots,\mathbf{x}_p))=\|\mathbf{x}_1\wedge\cdots\mathbf{x}_p\|\int_U du_1\cdots du_p. \tag{7.56}$$

■

Although the formula of a triangle's area is well known, "half the base times the height", much less known, however, is that this formula can be easily generalized for a simplex in a dimension p:

PROPOSITION 7.3.– Let $\mathbf{x}_1,\cdots,\mathbf{x}_p$ be a linearly independent family of p vectors of $\boldsymbol{E}$ such that $\mathbf{x}_p$ is orthogonal to each among $\mathbf{x}_1,\cdots,\mathbf{x}_{p-1}$. Then, we have the formula:

$$\mathcal{V}(\mathcal{S}(\mathbf{x}_1\cdots\mathbf{x}_p))=\frac{1}{p}\mathcal{V}(\mathcal{S}(\mathbf{x}_1\cdots\mathbf{x}_{p-1}))\|\mathbf{x}_p\|. \tag{7.57}$$

In other words, for a simplex with a dimension p, we have: "measure equals base times height divided by p".

PROOF.– First, note that:

$$\|\mathbf{x}_1\wedge\cdots\wedge\mathbf{x}_p\|^2=Q_p(\mathbf{x}_1\wedge\cdots\wedge\mathbf{x}_p). \tag{7.58}$$

By developing the right-hand side using the definition of the dot product:

$$Q_p(\mathbf{x}_1\wedge\cdots\wedge\mathbf{x}_p)=\det(b(\mathbf{x}_i,\mathbf{x}_j)). \tag{7.59}$$

Since $\mathbf{x}_p$ is orthogonal to all the $x_i, i\in[1,p-1]$, the last column of the matrix, whose determinant has to be calculated, is such that all its coefficients are null, with exception of the last one which equals $q(\mathbf{x}_p)$. By developing the determinant with respect to the last column, we then immediately have:

$$Q_p(\mathbf{x}_1\wedge\cdots\wedge\mathbf{x}_p)=Q_{p-1}(\mathbf{x}_1\wedge\cdots\wedge\mathbf{x}_{p-1})\,q(\mathbf{x}_p), \tag{7.60}$$

hence, we deduce:

$$\|\mathbf{x}_1\wedge\cdots\wedge\mathbf{x}_p\|=\|\mathbf{x}_1\wedge\cdots\wedge\mathbf{x}_{p-1}\|\,\|\mathbf{x}_p\|, \tag{7.61}$$

finally, we easily obtain:

$$\mathcal{V}(\mathcal{S}(\mathbf{x}_1,\cdots,\mathbf{x}_p)) = \frac{1}{p!}\|\mathbf{x}_1\wedge\cdots\wedge\mathbf{x}_{p-1}\|\,\|\mathbf{x}_p\|$$

$$= \frac{1}{p}\frac{1}{(p-1)!}\|\mathbf{x}_1\wedge\cdots\wedge\mathbf{x}_{p-1}\|\,\|\mathbf{x}_p\| = \frac{\|\mathbf{x}_p\|}{p}\mathcal{V}(\mathcal{S}(\mathbf{x}_1\cdots\mathbf{x}_{p-1})), \quad [7.62]$$

and the result is demonstrated.

7.2.3. *Usage of coordinates of the exterior product vector*

By employing the coordinates of the vector $\mathbf{x}_1\wedge\cdots\wedge\mathbf{x}_p$ with respect to the orthonormal basis $\mathbf{e}_{i_1}\wedge\cdots\wedge\mathbf{e}_{i_p}$, we deduce the following corollary:

COROLLARY 7.5.– Let $\mathbf{M}$ be a matrix with a size (n,p), with $n\geq p$. In order to calculate the determinant of the matrix ${}^t\mathbf{MM}$ (squared, with a size p), we can proceed as follows:

1) create all the square matrices with a size p by removing the corresponding rows;

2) calculate the determinant for each matrix;

3) add up the squares of the obtained determinants.

PROOF.– By using the p columns of the matrix $\mathbf{M}$, we create the vectors $\mathbf{x}_1,\cdots,\mathbf{x}_p$ such that:

$$\mathbf{x}_j = \sum_{i=1}^{i=n} M_{ij}\mathbf{e}_i. \quad [7.63]$$

Then, we know that the determinant of the matrix ${}^t\mathbf{MM}$ is given by the norm square of the vector $\mathbf{x}_1\wedge\cdots\wedge\mathbf{x}_p$. We can calculate the components of this vector within the orthonormal basis $\mathbf{e}_{i_1}\wedge\cdots\wedge\mathbf{e}_{i_p}, \mathbf{i}\in\mathcal{I}_o$:

$$\mathbf{x}_1\wedge\cdots\wedge\mathbf{x}_p = \sum_{\mathbf{i}\in\mathcal{I}_o}\det(\tilde{\mathbf{x}}_1(\mathbf{i}),\cdots,\tilde{\mathbf{x}}_p(\mathbf{i}))\,\mathbf{e}_{i_1}\wedge\cdots\wedge\mathbf{e}_{i_p}. \quad [7.64]$$

Now, we can observe that we actually have:

$$\tilde{\mathbf{x}}_j(\mathbf{i}) = \sum_{l=1}^{l=p} x_{j,i_l}\mathbf{e}_{i_l} = \sum_{l=1}^{l=p} M_{i_l,j}\mathbf{e}_{i_l}. \quad [7.65]$$

Therefore, in order to construct the determinant of the family $\tilde{\mathbf{x}}_1(\mathbf{i}), \cdots, \tilde{\mathbf{x}}_p(\mathbf{i})$, we calculate the determinant of the square matrix with only the rows $i_1, \cdots, i_p$ of the matrix M. Since the basis $\mathbf{e}_{i_1} \wedge \cdots \wedge \mathbf{e}_{i_p}, \mathbf{i} \in \mathcal{I}_o$ is orthonormal, the square of the norm of the vector $\mathbf{x}_1 \wedge \cdots \wedge \mathbf{x}_p$ is obtained as a sum of the squares of its components. Hence our result.

7.2.4. *"Infinitesimal" measure of classical objects*

In this section, we provide a classical example of Riemann's measure: the measure of a sphere. For this purpose, we write "approximately" that we have:

$$M \in \mathcal{S} \Leftrightarrow \exists (\theta, \phi) \in [0, 2\pi] \times [0, \pi], \begin{cases} x_1 = r\cos(\theta)\sin(\phi) \\ x_2 = r\sin(\theta)\sin(\phi) \\ x_3 = r\cos(\phi) \end{cases}. \quad [7.66]$$

The matrix of tangent vectors $\mathbf{M}$ is a matrix with three rows and two columns, expressed as:

$$\mathbf{M} = \begin{bmatrix} -r\sin(\theta)\sin(\phi) & r\cos(\theta)\cos(\phi) \\ r\cos(\theta)\sin(\phi) & r\sin(\theta)\cos(\phi) \\ 0 & -r\sin(\phi) \end{bmatrix}. \quad [7.67]$$

Then, we can proceed in two ways:

1) We calculate the matrix ${}^tMM$. It equals:

$$\begin{bmatrix} r^2\sin^2(\phi) & 0 \\ 0 & r^2 \end{bmatrix} \quad [7.68]$$

hence, we deduce $\sqrt{\det({}^tMM)} = r^2|\sin(\phi)| = r^2\sin(\phi)$ and the associated measure $r^2\sin(\phi)\,d\theta d\phi$.

2) We calculate at first the components of the column vectors $M_{:,1} \wedge M_{:,2}$ with respect to the basis $\mathbf{e}_i \wedge \mathbf{e}_j, i < j$. We obtain thus:

$$M_{:,1} \wedge M_{:,2} = -r^2\sin(\phi)\cos(\phi)\,\mathbf{e}_1 \wedge \mathbf{e}_2 + r^2\sin(\theta)\sin^2(\phi)\,\mathbf{e}_1 \wedge \mathbf{e}_3 - r^2\cos(\theta)\sin^2(\phi)\,\mathbf{e}_2 \wedge \mathbf{e}_3, \quad [7.69]$$

then, we consider the Euclidean norm of $M_{:,1} \wedge M_{:,2}$; in other words, we calculate:

$$Q_2^+(M_{:,1} \wedge M_{:,2}) = r^4\sin^2(\phi)\cos^2(\phi) + r^4\sin^2(\theta)\sin^4(\phi) + r^4\cos^2(\theta)\sin^4(\phi) = r^4\sin^2(\phi) \quad [7.70]$$

with an identical result as before (of course...). One of the characteristics of the second method is that it has more steps, but each step has simpler calculations. A direct calculation of ${}^{t}MM$ might be more difficult to write (for students...) than the direct calculation of exterior products.

8

Divisibility and Decomposability

Let us consider an algebra $\Lambda^p E$, where E is a linear space with a dimension n. We know that all the elements of the form $\mathbf{x}_1 \wedge \cdots \wedge \mathbf{x}_p$ belong to $\Lambda^p E$. However, we also know that there exist vectors of $\Lambda^p E$ which cannot be written in this form. Therefore, the following issue arises:

QUESTION 8.1.– Let E be a space with a dimension n and $\mathbf{E}$ an element of $\Lambda^p E$. Does there exist a method for characterizing whether this element can be written as an exterior product of p vectors of E?

More generally, since we have a symbol of product between elements of exterior algebras, we can naturally consider all the issues of divisibility by elements of exterior algebras.

8.1. Contraction product

8.1.1. *Definition*

The divisibility issue requires the implementation of several rather useful calculation tools. Let us begin with the definition of the contraction product between elements of the exterior algebras:

DEFINITION 8.1.– Let $\mathbf{X} \in \Lambda^p E$ and $\mathbf{Y} \in \Lambda^q E$ with $p \geq q$. Then, there exists a unique element of $\Lambda^{p-q} E$, denoted by $\mathbf{X} : \mathbf{Y}$, that satisfies:

$$\forall \mathbf{Z} \in \Lambda^{p-q} E, \quad B_{p-q}(\mathbf{X} : \mathbf{Y}, \mathbf{Z}) = B_p(\mathbf{X}, \mathbf{Z} \wedge \mathbf{Y}). \qquad [8.1]$$

PROOF.– This is an immediate application of the Riesz representation theorem 6.2: for some fixed $\mathbf{X}, \mathbf{Y}$, the map $\mathbf{Z} \mapsto B_p(\mathbf{X}, \mathbf{Z} \wedge \mathbf{Y})$ is a linear form over $\Lambda^{p-q}E$. Therefore, there exists a unique vector of $\Lambda^{p-q}E$ that represents this form (via the action of the pseudo-dot product B_p). This is precisely the vector in the theorem.

REMARK 8.1.– The following points should be mentioned:

1) When $q = 0$, i.e. when $\mathbf{Y} = y$ is a real, we have $\mathbf{X} : y = y\mathbf{X}$ in the sense of a product of a scalar by a vector. This is due to the fact that for the reals, we have set by definition 3.1 $\mathbf{Z} \wedge y = y \wedge \mathbf{Z} = y\mathbf{Z}$ in the sense of a product of a scalar by a vector;

2) When $p = q$, the contraction product becomes a pseudo-dot product B_p between p-vectors. ■

A property, which we shall not discuss in detail, is straightforward:

PROPOSITION 8.1.– The contraction symbol ":" is a bilinear symbol with respect to its arguments:

$$\forall \mathbf{X}, \mathbf{X}', \forall \mathbf{Y}, \forall \alpha \in \mathbb{R}, \quad (\mathbf{X} + \lambda \mathbf{X}') : \mathbf{Y} = \mathbf{X} : \mathbf{Y} + \alpha \mathbf{X}' : \mathbf{Y} \qquad [8.2]$$

$$\forall \mathbf{X}, \forall \mathbf{Y}, \mathbf{Y}', \forall \beta \in \mathbb{R}, \quad \mathbf{X} : (\mathbf{Y} + \beta \mathbf{Y}') = \mathbf{X} : \mathbf{Y} + \beta \mathbf{X} : \mathbf{Y}' \qquad [8.3]$$

8.1.2. *Calculation formulas*

An interesting question concerns the action of a p-blade (on the left) upon a vector (on the right). In fact, in its explicit formulation, it is related to the expansion of a determinant by a row or a column. For that matter, it is precisely the Laplace expansion that allows us to demonstrate the following formula:

PROPOSITION 8.2.– Let $\mathbf{E} := \mathbf{x}_1 \wedge \cdots \wedge \mathbf{x}_p$ be a p-blade. Then, we have:

$$\mathbf{E} : \mathbf{x} = \sum_{k=1}^{k=p} (-1)^{p+k} b(\mathbf{x}, \mathbf{x}_k) (\mathbf{x}_1 \wedge \cdots \wedge \mathbf{x}_p / [\mathbf{x}_k]). \qquad [8.4]$$

PROOF.– Let $\mathbf{Y} := \mathbf{y}_1 \wedge \cdots \wedge \mathbf{y}_p$ be a $p-1$ blade of $\Lambda^{p-1}E$, then necessarily:

$$B_{p-1}(\mathbf{E} : \mathbf{x}, \mathbf{Y}) = B_p(\mathbf{E}, \mathbf{Y} \wedge \mathbf{x}) \Leftrightarrow B_{p-1}(\mathbf{E} : \mathbf{x}, \mathbf{y}_1 \wedge \cdots \wedge \mathbf{y}_{p-1})$$
$$= B_p(\mathbf{x}_1 \wedge \cdots \mathbf{x}_p, \mathbf{y}_1 \wedge \cdots \wedge \mathbf{y}_{p-1} \wedge \mathbf{x}), \qquad [8.5]$$

which yields:

$$B_{p-1}(\mathbf{E}:\mathbf{x},\mathbf{Y}) = B_p(\mathbf{x}_1 \wedge \cdots \mathbf{x}_p, \mathbf{y}_1 \wedge \cdots \wedge \mathbf{y}_{p-1} \wedge \mathbf{x}) = \det(A), \qquad [8.6]$$

with A_{ij} being the matrix constructed over the dot products $\mathbf{x}_i \cdot \mathbf{y}_j$ (thus we shall set $\mathbf{x} := \mathbf{y}_p$). By expanding this determinant with respect to the last column (see Laplace's formula 5.3), we quickly obtain:

$$B_{p-1}(\mathbf{E}:\mathbf{x},\mathbf{Y}) = \sum_{k=1}^{k=p} b(\mathbf{x},\mathbf{x}_k)(-1)^{p+k}\det\left(\tilde{A}(k,p)\right), \qquad [8.7]$$

where the matrix $\tilde{A}(k,p)$ is obtained from the matrix A by removing the row k and the column p. Such a determinant can be written as a dot product between the vector $(\mathbf{x}_1 \wedge \cdots \wedge \mathbf{x}_p / [\mathbf{x}_k])$ and the vector $\mathbf{y}_1 \wedge \cdots \wedge \mathbf{y}_{p-1}$. Therefore, we have:

$$B_{p-1}(\mathbf{E}:\mathbf{x},\mathbf{Y}) = \sum_{k=1}^{k=p} b(\mathbf{x},\mathbf{x}_k)\, B_{p-1}(\mathbf{x}_1 \wedge \cdots \wedge \mathbf{x}_p / [\mathbf{x}_k], \mathbf{Y}). \qquad [8.8]$$

Hence, by identification, we deduce the result stated by the proposition:

$$(\mathbf{x}_1 \wedge \cdots \wedge \mathbf{x}_p) : \mathbf{x} = \sum_{k=1}^{k=p} (-1)^{p+k} b(\mathbf{x},\mathbf{x}_k)(\mathbf{x}_1 \wedge \cdots \wedge \mathbf{x}_p / [\mathbf{x}_k]). \qquad [8.9]$$

PROPOSITION 8.3.– Let $\mathbf{H} := \mathbf{x}_1 \wedge \cdots \wedge \mathbf{x}_p$ be a non-null p-blade. We denote by $\boldsymbol{H}$ the subspace of dimension p generated by the (linearly independent, see Proposition 3.8) family $\mathbf{x}_1, \cdots, \mathbf{x}_p$. Then, the following affirmations hold:

1) we have $\mathbf{x} \in \boldsymbol{H} \Leftrightarrow \mathbf{H} \wedge \mathbf{x} = \mathbf{0}$;

2) we have $\mathbf{x} \in \boldsymbol{H}^{\perp} \Leftrightarrow \mathbf{H} : \mathbf{x} = \mathbf{0}$.

PROOF.– The vector $\mathbf{x}$ belongs to the subspace $\boldsymbol{H}$ if and only if the family $\mathbf{x}_1, \cdots, \mathbf{x}_p, \mathbf{x}$ is linearly dependent and thus if and only if $\mathbf{H} \wedge \mathbf{x} = \mathbf{0}$ (see Proposition 3.9). Similarly, we have $\mathbf{x} \in \boldsymbol{H}^{\perp}$ if and only if $\forall k \in [1,p]; b(\mathbf{x},\mathbf{x}_k) = 0$. Since the family $\mathbf{x}_1, \cdots, \mathbf{x}_p$ is linearly independent, the set of

$$(\mathbf{x}_1 \wedge \cdots \wedge \mathbf{x}_p / [\mathbf{x}_k]),\ k \in [1,p] \qquad [8.10]$$

is a basis of $\Lambda^{p-1}\boldsymbol{H}$. As a result, we have $\forall k \in [1,p]$, $b(\mathbf{x},\mathbf{x}_k) = 0$ if and only if:

$$\sum_{k=1}^{k=p} (-1)^{p+k} b(\mathbf{x},\mathbf{x}_k)(\mathbf{x}_1 \wedge \cdots \wedge \mathbf{x}_p / [\mathbf{x}_k]) = \mathbf{0}, \qquad [8.11]$$

which represents precisely $\mathbf{H} : \mathbf{x} = \mathbf{0}$.

If $\mathbf{H}$ is a non-null p-blade, we cannot simultaneously have $\mathbf{H} \wedge \mathbf{x} = \mathbf{0}$ and $\mathbf{H} : \mathbf{x} = \mathbf{0}$ for $q(\mathbf{x}) \neq \mathbf{0}$ as it would mean that $\mathbf{x}$ belongs simultaneously to $\boldsymbol{H}$ and to $\boldsymbol{H}^{\perp}$, which is impossible since we would have $q(\mathbf{x}) = 0$, and this condition is excluded by the hypothesis. We obtain the following characterization, whose importance should not be underestimated:

PROPOSITION 8.4.– Let $\mathbf{E} \in \Lambda^p \boldsymbol{E}$. Then, $\mathbf{E} = \mathbf{0}$ if and only if there exists $\mathbf{x} \in \boldsymbol{E}$, $q(\mathbf{x}) \neq 0$ such that we have simultaneously:

$$\mathbf{E} : \mathbf{x} = \mathbf{0} \textit{ and } \mathbf{E} \wedge \mathbf{x} = \mathbf{0}. \quad [8.12]$$

PROOF.– Actually, we can always suppose that $\mathbf{x}$ has a unitary (pseudo-) norm. We denote it by $\mathbf{e}_1$ and we complete this vector by a pseudo-orthonormal basis $\mathbf{e}_2, \cdots, \mathbf{e}_n$ (see corollary 6.4). We then write $\mathbf{E}$ in the form:

$$\mathbf{E} = \sum_{\mathrm{i}} \theta_{\mathrm{i}} \mathbf{e}_{i_1} \wedge \cdots \wedge \mathbf{e}_{i_p}. \quad [8.13]$$

By exploiting the fact that:

$$\mathbf{E} \wedge \mathbf{e}_1 = \mathbf{0} \Leftrightarrow \left(\sum_{\mathrm{i}} \theta_{\mathrm{i}} \mathbf{e}_{i_1} \wedge \cdots \wedge \mathbf{e}_{i_p} \right) \wedge \mathbf{e}_1 = \mathbf{0}, \quad [8.14]$$

we reach the conclusion that all the elements θ_{i} with $i_1 > 1$ are null. Therefore, it remains only:

$$\mathbf{E} = \sum_{2 \leq i_2 < \cdots < i_p \leq n} \theta_{1 i_2 \cdots i_p} \mathbf{e}_1 \wedge \cdots \mathbf{e}_{i_2} \wedge \cdots \wedge \mathbf{e}_{i_p}. \quad [8.15]$$

Then, we exploit the fact that:

$$\mathbf{E} : \mathbf{e}_1 = \mathbf{0} \Leftrightarrow \left(\sum_{2 \leq i_2 < \cdots < i_p \leq n} \theta_{1 i_2 \cdots i_p} \mathbf{e}_1 \wedge \cdots \mathbf{e}_{i_2} \wedge \cdots \wedge \mathbf{e}_{i_p} \right) : \mathbf{e}_1 = \mathbf{0}. \quad [8.16]$$

By using the explicit form of the action of a blade over a vector in a contraction product 8.2, we deduce thereof the conclusion that all the $\theta_{1 i_2 \cdots i_p}$ are null since $q(\mathbf{e}_1) \neq 0$ and that $b(\mathbf{e}_1, \mathbf{e}_i) = 0$ for $i > 1$.

The following is an interesting calculation rule:

PROPOSITION 8.5.– Let $\mathbf{E} \in \Lambda^p \boldsymbol{E}$ and let $\mathbf{x}_1, \cdots, \mathbf{x}_r$ (with $r \leq p$) be a sequence of elements of $\boldsymbol{E}$. Then, we have:

$$(\cdots((\mathbf{E} : \mathbf{x}_1) : \mathbf{x}_2) : \cdots) : \mathbf{x}_r = \mathbf{E} : (\mathbf{x}_r \wedge \cdots \wedge \mathbf{x}_1). \quad [8.17]$$

PROOF.– It suffices to note that:

$$\langle(\cdots((\mathbf{E}:\mathbf{x}_1):\mathbf{x}_2):\cdots):\mathbf{x}_r,\mathbf{Y}\rangle = \langle\mathbf{E},(\cdots((\mathbf{Y}\wedge\mathbf{x}_r)\wedge\mathbf{x}_{r-1})\wedge\cdots)\wedge\mathbf{x}_1\rangle. \quad [8.18]$$

Then, we employ the associativity of the exterior product to obtain:

$$\langle(\cdots((\mathbf{E}:\mathbf{x}_1):\mathbf{x}_2):\cdots):\mathbf{x}_r,\mathbf{Y}\rangle = \langle\mathbf{E},\mathbf{Y}\wedge(\mathbf{x}_r\wedge\cdots\wedge\mathbf{x}_1)\rangle. \quad [8.19]$$

The exterior products are moved to the left-hand side by using the definition of contraction product:

$$\langle(\cdots((\mathbf{E}:\mathbf{x}_1):\mathbf{x}_2):\cdots):\mathbf{x}_r,\mathbf{Y}\rangle = \langle\mathbf{E}:(\mathbf{x}_r\wedge\cdots\wedge\mathbf{x}_1),\mathbf{Y}\rangle, \quad [8.20]$$

and we conclude by identifying the terms on the left: we have obtained precisely the result stated above.

A corollary follows immediately:

COROLLARY 8.1.– For all $\mathbf{E}\in\Lambda^p E$ (with $p\geq 2$) and for all $\mathbf{x}\in E$, we have:

$$(\mathbf{E}:\mathbf{x}):\mathbf{x}=\mathbf{0}. \quad [8.21]$$

PROOF.– $(\mathbf{E}:\mathbf{x}):\mathbf{x}=\mathbf{E}:(\mathbf{x}\wedge\mathbf{x})$ (according to the previous proposition), and thus it is null since $\mathbf{x}\wedge\mathbf{x}$ is null.

8.2. Divisibility by a k-blade

8.2.1. *General case*

Let us recall a rather clear notion of divisibility:

DEFINITION 8.2.– Let $p\geq q$. We say that $\mathbf{B}\in\Lambda^q E$ is a (left) divisor of $\mathbf{A}\in\Lambda^p E$ if there exists $\mathbf{C}\in\Lambda^{p-q}E$ such that it is possible to write:

$$\mathbf{A}=\mathbf{B}\wedge\mathbf{C}. \quad [8.22]$$

REMARK 8.2.– Two evident remarks follow:

1) In virtue of Formula 3.16, left divisibility entails right divisibility;

2) The divisibility issue is not considered here from a general point of view. Rather, we shall address a particular (left) divisibility, which consists of divisibility by vectors and k-blades. ■

Considering an element $\mathbf{E} \in \Lambda^p \boldsymbol{E}$, an important tool that helps study the divisibility by k-blades is the map operated by $\mathbf{E}$ (on the left), denoted by $\mathbf{E} \wedge \cdot$, and defined as:

$$\mathbf{x} \in \boldsymbol{E} \mapsto \mathbf{E} \wedge \mathbf{x} \in \Lambda^{p+1} \boldsymbol{E}. \qquad [8.23]$$

More specifically, we have the following theorem:

THEOREM 8.1.– Let $\mathbf{E} \in \Lambda^p \boldsymbol{E}$ be a non-null p-vector. Then, $\mathbf{E}$ is divisible by a k-blade $\mathbf{x}_1 \wedge \cdots \wedge \mathbf{x}_k$ if and only if $\mathbf{x}_1, \cdots, \mathbf{x}_k$ is a linearly independent family of $\ker(\mathbf{E} \wedge \cdot)$.

PROOF.– Suppose that it is possible to write $\mathbf{E} = (\mathbf{x}_1 \wedge \cdots \wedge \mathbf{x}_k) \wedge \mathbf{F}$ with $\mathbf{F} \in \Lambda^{p-k} \boldsymbol{E}$. Then, since $\mathbf{E}$ is non-null, the family $\mathbf{x}_1, \cdots, \mathbf{x}_k$ is clearly linearly independent (otherwise, we would have $\mathbf{x}_1 \wedge \cdots \wedge \mathbf{x}_k = \mathbf{0}$ and thus $\mathbf{E}$ would be null). Moreover, it is evident that:

$$\forall i \in [1, k], \; \mathbf{E} \wedge \mathbf{x}_i = (\mathbf{x}_1 \wedge \cdots \wedge \mathbf{x}_k) \wedge \mathbf{F} \wedge \mathbf{x}_i = \mathbf{0}, \qquad [8.24]$$

which proves that all the $\mathbf{x}_i$ indeed belong to $\ker(\mathbf{E} \wedge \cdot)$. Vice versa, let us now suppose that $\mathbf{x}_1, \cdots, \mathbf{x}_k$ is a linearly independent family of $\ker(\mathbf{E} \wedge \cdot)$. We complete this linearly independent family with vectors $\mathbf{x}_{k+1}, \cdots, \mathbf{x}_n$ to obtain a basis of $\boldsymbol{E}$. We know that it is possible to write $\mathbf{E}$ in the form:

$$\mathbf{E} = \sum_{1 \le i_1 < \cdots < i_p \le n} \theta_{i_1 \cdots i_p} \mathbf{x}_{i_1} \wedge \cdots \wedge \mathbf{x}_{i_p}. \qquad [8.25]$$

We write that $\mathbf{E} \wedge \mathbf{x}_1 = \mathbf{0}$, in other words, that:

$$\left[\sum_{1 \le i_1 < \cdots < i_p \le n} \theta_{i_1 \cdots i_p} \mathbf{x}_{i_1} \wedge \cdots \wedge \mathbf{x}_{i_p} \right] \wedge \mathbf{x}_1 = \mathbf{0}. \qquad [8.26]$$

This entails immediately that all the $\theta_{i_1 \cdots i_p}$ (the multi-indexes are ordered) with $i_1 > 1$ are null. Therefore, it remains:

$$\mathbf{E} = \mathbf{x}_1 \wedge \left(\sum_{2 \le i_2 < \cdots < i_p \le n} \theta_{1 i_2 \cdots i_p} \mathbf{x}_{i_2} \wedge \cdots \wedge \mathbf{x}_{i_p} \right). \qquad [8.27]$$

At this point, we consider the exterior product between $\mathbf{E}$ and $\mathbf{x}_2$, which has to be null. In the same way as before, we obtain that all the $\theta_{1i_2\cdots i_p}$ with $i_2 > 2$ are null. By repeating the reasoning for $\mathbf{x}_3, \cdots, \mathbf{x}_k$, we obtain that all the coefficients with $i_k > k$ are null. Therefore, it remains only:

$$\mathbf{E} = \mathbf{x}_1 \wedge \cdots \wedge \mathbf{x}_k \wedge \left(\sum_{i_{k+1}<\cdots<i_p} \theta_{1\cdots k i_{k+1}\cdots i_p} \mathbf{x}_{i_{k+1}} \wedge \cdots \wedge \mathbf{x}_{i_p} \right), \qquad [8.28]$$

hence our result is deduced.

8.2.2. *Exterior vector division*

An immediate corollary regarding the problem of vector division follows:

COROLLARY 8.2.– Let $\mathbf{E} \in \Lambda^{p+1}E$ be non-null and let $\mathbf{a} \in E$ be non-null. The problem of exterior right division

$$\textit{Find } \mathbf{X} \textit{ within } \Lambda^p E \textit{ such that } : \mathbf{a} \wedge \mathbf{X} = \mathbf{E} \qquad [8.29]$$

admits a solution if and only if $\mathbf{E} \wedge \mathbf{a} = \mathbf{0}$.

PROOF.– Actually, it is possible to find $\mathbf{X}$ as specified if and only if the 1-blade $\mathbf{a}$ (i.e. the vector $\mathbf{a}$) is a left divisor of $\mathbf{E}$, and thus if and only if $\mathbf{a}$ is a non-null vector of $\ker(\mathbf{E} \wedge \cdot)$, and therefore if and only if $\mathbf{E} \wedge \mathbf{a} = \mathbf{0}$.

We now pose the following question: how to find a solution of the exterior division problem "explicitly". We have the following result:

THEOREM 8.2.– Let $\mathbf{E} \neq \mathbf{0} \in \Lambda^p E$ and $\mathbf{a} \in E$, $q(\mathbf{a}) \neq 0$ such that $\mathbf{E} \wedge \mathbf{a} = \mathbf{0}$. Then:

1) the equation $\mathbf{a} \wedge \mathbf{X} = \mathbf{E}$ admits a *unique* solution $\mathbf{S}$ such that $\mathbf{S} : \mathbf{a} = \mathbf{0}$;

2) this solution $\mathbf{S}$ is calculated as:

$$\mathbf{S} = (-1)^{p+1} \mathbf{E} : \left[\frac{\mathbf{a}}{q(\mathbf{a})} \right]. \qquad [8.30]$$

PROOF.– Regarding uniqueness, let us suppose that $\mathbf{S}_1, \mathbf{S}_2$ are two solutions that satisfy $\mathbf{S}_1 : \mathbf{a} = \mathbf{S}_2 : \mathbf{a} = \mathbf{0}$. Since by definition they satisfy $\mathbf{S}_1 \wedge \mathbf{a} = \mathbf{S}_2 \wedge \mathbf{a} = \mathbf{E}$, we can write simultaneously:

$$(\mathbf{S}_1 - \mathbf{S}_2) \wedge \mathbf{a} = \mathbf{0} \textit{ and } (\mathbf{S}_1 - \mathbf{S}_2) : \mathbf{a} = \mathbf{0} \qquad [8.31]$$

and since $q(\mathbf{a}) \neq 0$, this implies that $\mathbf{S}_1 - \mathbf{S}_2 = \mathbf{0}$ according to Proposition 8.4. Meanwhile, regarding the existence of such a solution, since we have $\mathbf{E} \wedge \mathbf{a} = \mathbf{0}$, we know that it is possible to write:

$$\mathbf{E} = \mathbf{a} \wedge \left(\sum_{2 \leq i_2 < \cdots < i_p \leq n} \eta_{i_2 \cdots i_p} \mathbf{e}_{i_2} \wedge \cdots \wedge \mathbf{e}_{i_p} \right), \qquad [8.32]$$

where each $\mathbf{e}_i, 2 \leq i$ is orthogonal to $\mathbf{a}$ (since $q(\mathbf{a}) \neq 0$ it is, in fact, possible to complete $\mathbf{a}$ by some vectors to obtain an orthogonal basis, see Corollary 6.4). We then calculate:

$$\mathbf{E} : \left[\frac{\mathbf{a}}{q(\mathbf{a})} \right] = \left(\mathbf{a} \wedge \sum_{2 \leq i_2 < \cdots < i_p \leq n} \eta_{i_2 \cdots i_p} \mathbf{e}_{i_2} \wedge \cdots \wedge \mathbf{e}_{i_p} \right) : \left[\frac{\mathbf{a}}{q(\mathbf{a})} \right], \qquad [8.33]$$

in other words,

$$\mathbf{E} : \left[\frac{\mathbf{a}}{q(\mathbf{a})} \right] = \sum_{2 \leq i_2 < \cdots < i_p \leq n} \eta_{i_2 \cdots i_p} \left[\mathbf{a} \wedge \mathbf{e}_{i_2} \wedge \cdots \wedge \mathbf{e}_{i_p} \right] : \left[\frac{\mathbf{a}}{q(\mathbf{a})} \right]. \qquad [8.34]$$

According to Formula 8.2, by taking into account that $\mathbf{a}$ is orthogonal to each $\mathbf{e}_{i_k}$, we have:

$$\left[\mathbf{a} \wedge \mathbf{e}_{i_2} \wedge \cdots \wedge \mathbf{e}_{i_p} \right] : \left[\frac{\mathbf{a}}{q(\mathbf{a})} \right] = (-1)^{p+1} \mathbf{e}_{i_2} \wedge \cdots \wedge \mathbf{e}_{i_p}. \qquad [8.35]$$

Hence, we deduce:

$$\sum_{2 \leq i_2 < \cdots < i_p \leq n} \eta_{i_2 \cdots i_p} \mathbf{e}_{i_2} \wedge \cdots \wedge \mathbf{e}_{i_p} = (-1)^{p+1} \mathbf{E} : \frac{\mathbf{a}}{q(\mathbf{a})}. \qquad [8.36]$$

Since we know that:

$$\mathbf{a} \wedge \left(\sum_{2 \leq i_2 < \cdots < i_p \leq n} \eta_{i_2 \cdots i_p} \mathbf{e}_{i_2} \wedge \cdots \wedge \mathbf{e}_{i_p} \right) = \mathbf{E}, \qquad [8.37]$$

we see that the term

$$(-1)^{p+1} \mathbf{E} : \frac{\mathbf{a}}{q(\mathbf{a})} \qquad [8.38]$$

is thus a solution of the vector division problem. Furthermore, it satisfies:

$$\left[(-1)^{p+1}\mathbf{E} : \frac{\mathbf{a}}{q(\mathbf{a})}\right] : \mathbf{a} = \frac{(-1)^{p+1}}{q(\mathbf{a})}\mathbf{E} : (\mathbf{a} \wedge \mathbf{a}) = \mathbf{0}, \quad [8.39]$$

hence our result.

8.3. Decomposability

DEFINITION 8.3.– Let $\mathbf{E} \in \Lambda^p \boldsymbol{E}$ be a p-vector. We say that $\mathbf{E}$ is:

1) decomposable if and only if there exists a family of p vectors of $\boldsymbol{E}$ such that $\mathbf{E} = \mathbf{x}_1 \wedge \cdots \wedge \mathbf{x}_p$;

2) pseudo-orthodecomposable if there exists a pseudo-orthogonal family into which it can be decomposed.

A decomposable p-vector of $\Lambda^p \boldsymbol{E}$ is called a p-blade.

We have the following characterization:

PROPOSITION 8.6.– A p-vector $\mathbf{E} \in \Lambda^p \boldsymbol{E}$ is decomposable if and only if it is pseudo-orthodecomposable.

PROOF.– This is a direct consequence of the construction process of orthogonal bases. Suppose that the vector $\mathbf{E}$ can be written in the form:

$$\mathbf{E} = \mathbf{x}_1 \wedge \cdots \wedge \mathbf{x}_p. \quad [8.40]$$

Since $\mathbf{E}$ is non-null, the space $\boldsymbol{F} := \langle \mathbf{x}_1, \cdots, \mathbf{x}_p \rangle$ has a dimension p. Then, it is always possible to construct a pseudo-orthogonal basis for $\boldsymbol{F}$ (see Corollary 6.5). If $\mathbf{e}_1, \cdots, \mathbf{e}_p$ represents such a basis, we then have:

$$\mathbf{E} = \det_{\mathbf{e}_1, \cdots, \mathbf{e}_p} (\mathbf{x}_1, \cdots, \mathbf{x}_p)\, \mathbf{e}_1 \wedge \cdots \wedge \mathbf{e}_p, \quad [8.41]$$

and thus $\mathbf{E}$ is indeed pseudo-orthodecomposable.

An important case of decomposability consists of the following:

PROPOSITION 8.7.– Let $(\boldsymbol{E}, b)$ be a pseudo-Euclidean space with a dimension n. Then, any element of $\Lambda^{n-1} \boldsymbol{E}$ is decomposable.

PROOF.– Consider an element of $\mathbf{E} \in \Lambda^{n-1}\boldsymbol{E}$. Then, it can be written in the form:

$$\mathbf{E} = \sum_{i=1}^{i=n} \theta_i \left(\mathbf{e}_1 \wedge \cdots \wedge \mathbf{e}_n \backslash [\mathbf{e}_i]\right). \quad [8.42]$$

Since $\mathbf{E}$ is non-null, not all θ_i are null. Let $\mathbf{x} = \sum_i x_i \mathbf{e}_i$ be an element of $\boldsymbol{E}$. The kernel of the map $\mathbf{E} \wedge \cdot$ is characterized by the equation:

$$\mathbf{E} \wedge \mathbf{x} = \mathbf{0} \Leftrightarrow \sum_{i=1}^{i=n} \theta_i \left(\mathbf{e}_1 \wedge \cdots \wedge \mathbf{e}_n \backslash [\mathbf{e}_i]\right) \wedge \sum_{j=1}^{j=n} x_j \mathbf{e}_j = \mathbf{0}. \quad [8.43]$$

Clearly, this can be written as:

$$\sum_{i=1}^{i=n} \theta_i x_i \left(\mathbf{e}_1 \wedge \cdots \wedge \mathbf{e}_n \backslash [\mathbf{e}_i]\right) \wedge \mathbf{e}_i = \mathbf{0}. \quad [8.44]$$

Now, for all $i \in [1, n]$, we have the equality:

$$\left(\mathbf{e}_1 \wedge \cdots \wedge \mathbf{e}_n \backslash [\mathbf{e}_i]\right) \wedge \mathbf{e}_i = (-1)^{n-i} \mathbf{e}_1 \wedge \cdots \wedge \mathbf{e}_n. \quad [8.45]$$

Hence, we deduce the characteristic equation of the kernel of the map $\mathbf{E} \wedge \cdot$:

$$\sum_{i=1}^{i=n} (-1)^{n-i} \theta_i x_i = 0, \quad [8.46]$$

this is of course an equation of a hyperplane, since not all θ_i are null. Therefore, the dimension of the kernel is indeed equal to $n - 1$. We apply a consequence of theorem 8.1: a vector of $\Lambda^p \boldsymbol{E}$ is decomposable if and only if $\dim(\ker(\mathbf{E} \wedge \cdot)) = p$. We have thus demonstrated what was required.

9

H-conjugation and Regressive Product

The introduction of the Hodge conjugation is one of the most significant contributions of Grassmann's formalism. First of all, the indispensable research of such an operation starts from the following question, a question that any somewhat curious science student has posed himself or herself at least once during their studies:

QUESTION 9.1.– Is there a way of generalizing the vector cross product operation to spaces with a dimension different than 3?

It is quite remarkable that the answer to such a naive question lies in such a profound operation as the Hodge conjugation, which is, however, a rather useful tool, if not an indispensable one, within the formalism of differential geometry or theoretical physics.

9.1. Introduction to H-conjugation

9.1.1. *Definition*

DEFINITION 9.1.– Let $(\boldsymbol{E}, b)$ be a pseudo-Euclidean space with a dimension n and a signature s. Then, there exist two vectors of $(\Lambda^n \boldsymbol{E}, B_n)$ that are equal to $(-1)^s$ for the quadratic form Q_n. In general, we denote by $\mathbf{I}$ such an element (which is also called pseudovector or caliber). Let $p \in [0, n]$. We call Hodge conjugate (or H-conjugate) of a p-vector $\mathbf{E}$ with respect to $\mathbf{I}$ the $(n-p)$-vector, defined as:

$$\mathbf{I} : \mathbf{E}, \qquad [9.1]$$

such a conjugate element is denoted by $\overline{\mathbf{E}}$, $\overline{\mathbf{E}}\big| \mathbf{I}$, depending on the case.

REMARK 9.1.– Before proceeding further, let us highlight some remarks:

1) The conjugation term seems quite appropriate, in particular in relation to complex numbers. In fact, if $z \in \mathbb{C}$, its conjugate $\overline{z}$ is such that $z\overline{z} = |z|^2$. In particular, provided that $|z|^2 \neq 0$, it is possible to write the reciprocal of z in the form $z^{-1} = \overline{z}/|z|^2$. We shall see that this kind of consideration can be extended to p-blades. However, in common vocabulary, it often happens that an operation defined in this way is called duality (rather than conjugation). In that case, we denote the element by a star notation *, instead of an overline notation $\overline{()}$.

2) Sometimes, the definition of the Hodge conjugate is introduced in another way, in particular by relating it to the vector division, most often in Euclidean spaces (and not in pseudo-Euclidean ones). In such cases, the authors consider an orthonormal basis $\mathbf{e}_1, \cdots, \mathbf{e}_n$ such that $\mathbf{e}_1 \wedge \cdots \wedge \mathbf{e}_n = \mathbf{I}$, and they define the conjugate in the following way:

$$\overline{\mathbf{e}_1 \wedge \cdots \wedge \mathbf{e}_p} = \mathbf{e}_{p+1} \wedge \cdots \wedge \mathbf{e}_n, \qquad [9.2]$$

then, they extend the definition of the conjugate to all the p-vectors by means of the operator's linearity. The liability of introducing such a definition lies in the fact that it is unclear why it should be modified in the pseudo-Euclidean case, since the right-hand side is multiplied by $(-1)^s$, with s being the signature of the pseudo-dot product. The interest in defining the conjugate as a contraction product over the element $\mathbf{I}$ is that this definition remains the same in any case. Of course, we shall see that these definitions are equivalent in the Euclidean case.

In accordance to our suggestion related to complex numbers, the H-conjugation can be interpreted in terms of an "inversion". In order to understand this concept, the following theorem, a quite fundamental one, even beyond the conjugation topic, is introduced. It generalizes the study of the problem of exterior division from corollary 8.2, by considering not simply a given vector $\mathbf{a}$, but rather a p-blade $\mathbf{a}_1 \wedge \cdots \wedge \mathbf{a}_p$.

THEOREM 9.1.– Let $\mathbf{B} \in \Lambda^{p+r}\boldsymbol{E}$ (with p, r two natural integers). Let $\mathbf{a}_1 \wedge \cdots \wedge \mathbf{a}_p$ be an anisotropic p-blade, in other words that satisfies the hypothesis: $Q_p(\mathbf{a}_1 \wedge \cdots \wedge \mathbf{a}_p) \neq 0$. Then, the following statements are true:

1) The exterior equation $(\mathbf{a}_1 \wedge \cdots \wedge \mathbf{a}_p) \wedge \mathbf{X} = \mathbf{B}$ admits a solution in $\Lambda^r\boldsymbol{E}$ if and only if:

$$\forall i \in [1, p], \quad \mathbf{B} \wedge \mathbf{a}_i = \mathbf{0}. \qquad [9.3]$$

2) The $(n-p)$-vector calculated according to the formula:

$$(-1)^{pr}\,\mathbf{B}:\left[\frac{(\mathbf{a}_1\wedge\cdots\wedge\mathbf{a}_p)}{Q_p(\mathbf{a}_1\wedge\cdots\wedge\mathbf{a}_p)}\right] \qquad [9.4]$$

is a solution of the equation.

3) This solution $\mathbf{X}$ is the only solution of the equation, and it satisfies additionally:

$$\forall i\in[1,p],\ \ \mathbf{X}:\mathbf{a}_i=\mathbf{0}. \qquad [9.5]$$

PROOF.– Let us demonstrate these properties:

1) Since we have $Q_p(\mathbf{a}_1\wedge\cdots\wedge\mathbf{a}_p)\neq 0$, it means that the family $\mathbf{a}_1,\cdots,\mathbf{a}_p$ is linearly independent (see corollary 7.3). In particular, according to theorem 8.1, the p-blade $\mathbf{a}_1\wedge\cdots\wedge\mathbf{a}_p\neq\mathbf{0}$ divides $\mathbf{B}$ if and only if each $\mathbf{a}_i$ is an element of $\ker(\mathbf{B}\wedge\cdot)$, and thus if and only if we have $\forall i\in[1,p],\ \mathbf{B}\wedge\mathbf{a}_i=\mathbf{0}$.

2) We have to show that the proposed vector actually solves the equation. Since the space $\boldsymbol{F}$ generated by the vectors $\mathbf{a}_1,\cdots,\mathbf{a}_p$ is nondegenerate (since $Q_p(\mathbf{a}_1\wedge\cdots\wedge\mathbf{a}_p)\neq 0$, see corollary 7.2), it is possible to construct a pseudo-orthonormal basis $\mathbf{e}_1,\cdots,\mathbf{e}_p$ of $\boldsymbol{F}$ that also satisfies:

$$\mathbf{a}_1\wedge\cdots\wedge\mathbf{a}_p=\alpha\mathbf{e}_1\wedge\cdots\mathbf{e}_p,\ \ \alpha=\det_{\mathbf{e}_1,\cdots,\mathbf{e}_p}(\mathbf{x}_1,\cdots,\mathbf{x}_p). \qquad [9.6]$$

Furthermore, it is possible to complete the basis $\mathbf{e}_1,\cdots,\mathbf{e}_p$ with a basis $\mathbf{e}_{p+1},\cdots,\mathbf{e}_n$ to obtain a pseudo-orthonormal basis of $\boldsymbol{E}$. Let us write then:

$$\mathbf{B}=\sum_{i_1,\cdots,i_{p+r}}\beta_{i_1,\cdots i_{p+r}}\mathbf{e}_{i_1}\wedge\cdots\wedge\mathbf{e}_{i_{p+r}}. \qquad [9.7]$$

Note that it is possible to express the $\mathbf{e}_i$ as a function of $\mathbf{a}_j$ in the form:

$$\forall i\in[1,p],\ \ \mathbf{e}_i=\sum_{j=1}^{j=p}\alpha_{ij}\mathbf{a}_j, \qquad [9.8]$$

in particular, since we have $\mathbf{B}\wedge\mathbf{a}_j=\mathbf{0}$ for all the $j\in[1,p]$, we deduce thereof that we have indeed $\mathbf{e}_i\wedge\mathbf{B}=\mathbf{0}$ for all the $i\in[1,p]$, and this necessarily entails that (see the demonstration of theorem 8.1):

$$\mathbf{B}=\sum_{p+1\leq i_{p+1}<\cdots<i_{p+r}}\beta_{1\cdots p i_{p+1}\cdots i_{p+r}}\mathbf{e}_1\wedge\cdots\wedge\mathbf{e}_p\wedge\mathbf{e}_{i_{p+1}}\wedge\cdots\wedge\mathbf{e}_{i_{p+r}}. \qquad [9.9]$$

Now, let us calculate $\mathbf{B} : \mathbf{e}_1$ by decomposing explicitly $\mathbf{B}$ in the form:

$$\left[\sum_{p+1\leq i_{p+1}<\cdots<i_{p+r}} \beta_{1\cdots p i_{p+1}\cdots i_{p+r}} \mathbf{e}_1 \wedge \cdots \wedge \mathbf{e}_p \wedge \mathbf{e}_{i_{p+1}} \wedge \cdots \wedge \mathbf{e}_{i_{p+r}}\right] : \mathbf{e}_1, \qquad [9.10]$$

this can be written, thanks to the formula (see proposition 8.2 taking into account the orthogonality of the $\mathbf{e}_j, j \in [1,p]$):

$$(-1)^{p+r+1} q(\mathbf{e}_1) \times \sum_{p+1\leq i_{p+1}<\cdots<i_{p+r}} \beta_{1\cdots p i_{p+1}\cdots i_{p+r}} \mathbf{e}_2 \wedge \cdots \wedge \mathbf{e}_p \wedge \mathbf{e}_{i_{p+1}} \wedge \cdots \wedge \mathbf{e}_{i_{p+r}}, \qquad [9.11]$$

this process can be iterated to calculate $(\mathbf{B} : \mathbf{e}_1) : \mathbf{e}_2$ according to the formula:

$$(-1)^{p+r+1} (-1)^{p+r} q(\mathbf{e}_1) q(\mathbf{e}_2) \sum_{p+1\leq i_{p+1}<\cdots<i_{p+r}} \beta_{1\cdots p i_{p+1}\cdots i_{p+r}} \mathbf{e}_3 \wedge \cdots \wedge \mathbf{e}_p \wedge \mathbf{e}_{i_{p+1}} \wedge \cdots \wedge \mathbf{e}_{i_{p+r}}, \qquad [9.12]$$

to the extent that, in the end, it is possible to calculate $(\cdots((\mathbf{B} : \mathbf{e}_1) : \mathbf{e}_2) : \cdots) : \mathbf{e}_p$ via the following formula:

$$(-1)^{p+r+1} (-1)^{p+r} \cdots (-1)^{r+2} q(\mathbf{e}_1) \cdots q(\mathbf{e}_p) \sum_{p+1\leq i_{p+1}<\cdots<i_{p+r}} \beta_{1\cdots p i_{p+1}\cdots i_{p+r}} \mathbf{e}_{i_{p+1}} \wedge \cdots \wedge \mathbf{e}_{i_{p+r}}. \qquad [9.13]$$

Hence, according to proposition 8.5 regarding the sequences of contraction products, we deduce the equality:

$$\mathbf{B} : (\mathbf{e}_p \wedge \cdots \wedge \mathbf{e}_1) = (-1)^{n(p,r)} q(\mathbf{e}_1) \cdots q(\mathbf{e}_p) \sum_{p+1\leq i_{p+1}<\cdots<i_{p+r}} \beta_{1\cdots p i_{p+1}\cdots i_{p+r}} \mathbf{e}_{i_{p+1}} \wedge \cdots \wedge \mathbf{e}_{i_{p+r}}, \qquad [9.14]$$

where we have set:

$$n(p,r) = p(r+1) + \frac{p(p+1)}{2}. \qquad [9.15]$$

In particular, we obtain:

$$(-1)^{p(p-1)/2} \frac{\mathbf{B} : (\mathbf{e}_1 \wedge \cdots \wedge \mathbf{e}_p)}{q(\mathbf{e}_1) \cdots q(\mathbf{e}_p)} = (-1)^{n(p,r)} \sum_{p+1\leq i_{p+1}<\cdots<i_{p+r}} \beta_{1\cdots p i_{p+1}\cdots i_{p+r}} \mathbf{e}_{i_{p+1}} \wedge \cdots \wedge \mathbf{e}_{i_{p+r}}, \qquad [9.16]$$

in other words,

$$(-1)^{pr}\frac{\mathbf{B}:(\mathbf{e}_1\wedge\cdots\wedge\mathbf{e}_p)}{q(\mathbf{e}_1)\cdots q(\mathbf{e}_p)}=$$
$$\sum_{p+1\leq i_{p+1}<\cdots<i_{p+r}}\beta_{1\cdots p i_{p+1}\cdots i_{p+r}}\mathbf{e}_{i_{p+1}}\wedge\cdots\wedge\mathbf{e}_{i_{p+r}}.\qquad [9.17]$$

By composition on the left with $\mathbf{e}_1\wedge\cdots\wedge\mathbf{e}_p$, we thus obtain:

$$(\mathbf{e}_1\wedge\cdots\wedge\mathbf{e}_p)\wedge\left[(-1)^{pr}\frac{\mathbf{B}:(\mathbf{e}_1\wedge\cdots\wedge\mathbf{e}_p)}{q(\mathbf{e}_1)\cdots q(\mathbf{e}_p)}\right]=\mathbf{B}.\qquad [9.18]$$

By multiplying the numerator and the denominator by α^2 (which is non-null), we then obtain:

$$(\mathbf{a}_1\wedge\cdots\wedge\mathbf{a}_p)\wedge\left[(-1)^{pr}\frac{\mathbf{B}:(\mathbf{a}_1\wedge\cdots\wedge\mathbf{a}_p)}{Q_p(\mathbf{a}_1\wedge\cdots\wedge\mathbf{a}_p)}\right]=\mathbf{B},\qquad [9.19]$$

hence our result.

3) Now, it only remains to demonstrate the uniqueness of such a solution in the stated condition. Let $\mathbf{X}\in\Lambda^r\boldsymbol{E}$ be a vector that satisfies $\forall i\in[1,p],\mathbf{X}:\mathbf{e}_i=\mathbf{0}$ (with $p+r\leq n$). Then, necessarily, such a vector can be written as:

$$\mathbf{X}=\sum_{p+1\leq i_1<\cdots<i_r\leq n}X_{i_1\cdots i_r}\mathbf{e}_{i_1}\wedge\cdots\wedge\mathbf{e}_{i_r},\qquad [9.20]$$

however, if this vector satisfies additionally:

$$(\mathbf{e}_1\wedge\cdots\wedge\mathbf{e}_p)\wedge\mathbf{X}=\mathbf{0},\qquad [9.21]$$

then it is null, since we have:

$$(\mathbf{e}_1\wedge\cdots\wedge\mathbf{e}_p)\wedge\mathbf{X}=(\mathbf{e}_1\wedge\cdots\wedge\mathbf{e}_p)$$
$$\wedge\sum_{p+1\leq i_1<\cdots<i_r\leq n}X_{i_1\cdots i_r}\mathbf{e}_{i_1}\wedge\cdots\wedge\mathbf{e}_{i_r}=\mathbf{0}.\qquad [9.22]$$

The family

$$\mathbf{e}_1\wedge\cdots\wedge\mathbf{e}_p\wedge\mathbf{e}_{i_1}\wedge\cdots\wedge\mathbf{e}_{i_r},\quad p+1\leq i_1<\cdots<i_r\leq n\qquad [9.23]$$

is a linearly independent family of $\Lambda^{p+r}\boldsymbol{E}$. As a result, all the components

$$X_{i_1\cdots i_r},\quad p+1\leq i_1<\cdots<i_r\leq n\qquad [9.24]$$

are null and the vector $\mathbf{X}$ is null. Now, if there are two solutions $\mathbf{X}_1, \mathbf{X}_2$ that satisfy

$$(\mathbf{a}_1 \wedge \cdots \wedge \mathbf{a}_p) \wedge \{\mathbf{X}_1, \mathbf{X}_2\} = \mathbf{B}, \ \ \forall i \in [1,p], \ \{\mathbf{X}_1, \mathbf{X}_2\} : \mathbf{a}_i = \mathbf{0}, \qquad [9.25]$$

then the difference $\mathbf{X}_1 - \mathbf{X}_2$ satisfies

$$(\mathbf{a}_1 \wedge \cdots \wedge \mathbf{a}_p) \wedge (\mathbf{X}_1 - \mathbf{X}_2) = \mathbf{0}, \ \ \forall i \in [1,p], \ (\mathbf{X}_1 - \mathbf{X}_2) : \mathbf{a}_i = \mathbf{0}, \qquad [9.26]$$

since the $\mathbf{e}_j, j \in [1,p]$ are linear combinations of $\mathbf{a}_i$, and since $(\mathbf{a}_1 \wedge \cdots \wedge \mathbf{a}_p) = \alpha (\mathbf{a}_1 \wedge \cdots \wedge \mathbf{a}_p)$ with $\alpha \neq 0$, it is then possible to write:

$$(\mathbf{e}_1 \wedge \cdots \wedge \mathbf{e}_p) \wedge (\mathbf{X}_1 - \mathbf{X}_2) = \mathbf{0}, \ \ \forall i \in [1,p], \ (\mathbf{X}_1 - \mathbf{X}_2) : \mathbf{e}_i = \mathbf{0}, \qquad [9.27]$$

in particular, according to the previous argument, $\mathbf{X}_1 - \mathbf{X}_2$ is null.

It is now possible to exploit this divisibility theorem to make an interpretation of the Hodge conjugation in terms of an "inversion" with respect to the pseudovector $\mathbf{I}$:

PROPOSITION 9.1.– Let $\mathbf{a}_1, \cdots, \mathbf{a}_p$ be such that $Q_p(\mathbf{a}_1 \wedge \cdots \wedge \mathbf{a}_p) \neq 0$. Then, the Hodge conjugate of the p-blade $\mathbf{a}_1 \wedge \cdots \wedge \mathbf{a}_p$ acts as an inverse operation for $\mathbf{I}$, according to the formula:

$$(\mathbf{a}_1 \wedge \cdots \mathbf{a}_p) \wedge \overline{(\mathbf{a}_1 \wedge \cdots \mathbf{a}_p)} = (-1)^{p(n-p)} Q_p(\mathbf{a}_1 \wedge \cdots \mathbf{a}_p) \mathbf{I}. \qquad [9.28]$$

PROOF.– This is a simple consequence of the previous theorem: suppose that we have $Q_p(\mathbf{a}_1 \wedge \cdots \mathbf{a}_p) \neq 0$; since it is always true that:

$$\forall i \in [1,p], \ \mathbf{I} \wedge \mathbf{a}_i = \mathbf{0}, \qquad [9.29]$$

this means that the vector division problem $(\mathbf{a}_1 \wedge \cdots \mathbf{a}_p) \wedge \mathbf{X} = \mathbf{I}$ admits a unique solution that satisfies additionally $\forall i \in [1,p], \ \mathbf{X} : \mathbf{a}_i = \mathbf{0}$. This solution is given by the formula:

$$\mathbf{X} = (-1)^{p(n-p)} \frac{\mathbf{I} : (\mathbf{a}_1 \wedge \cdots \mathbf{a}_p)}{Q_p(\mathbf{a}_1 \wedge \cdots \mathbf{a}_p)} \qquad [9.30]$$

or, in other words,

$$\mathbf{X} = (-1)^{p(n-p)} \frac{\overline{\mathbf{a}_1 \wedge \cdots \wedge \mathbf{a}_p}}{Q_p(\mathbf{a}_1 \wedge \cdots \mathbf{a}_p)}. \qquad [9.31]$$

The interpretation presented above will enable a quick calculation of the H-conjugate of pseudo-orthonormal families:

COROLLARY 9.1.– Let $\mathbf{e}_1, \cdots, \mathbf{e}_p$ be a pseudo-orthonormal family. It is always possible to complete it with $\mathbf{e}_{p+1}, \cdots, \mathbf{e}_n$, another pseudo-orthonormal family, such that $\mathbf{e}_1 \wedge \cdots \wedge \mathbf{e}_n = \mathbf{I}$. Then, we have:

$$\overline{\mathbf{e}_1 \wedge \cdots \wedge \mathbf{e}_p} = (-1)^{p(n-p)} q(\mathbf{e}_1) \cdots q(\mathbf{e}_p) \mathbf{e}_{p+1} \wedge \cdots \wedge \mathbf{e}_n. \qquad [9.32]$$

PROOF.– We apply the previous result, by noting that for a pseudo-orthonormal family $\mathbf{e}_1, \cdots, \mathbf{e}_p$, the term $Q_p(\mathbf{e}_1 \wedge \cdots \wedge \mathbf{e}_p)$ is equal to $q(\mathbf{e}_1) \cdots q(\mathbf{e}_p)$.

To conclude this introduction, let us provide the following elements:

PROPOSITION 9.2.– Let $\boldsymbol{E}, b$ be a pseudo-Euclidean space with a signature s and a dimension n. Then, we have $\overline{1} = \mathbf{I}$ and $\overline{\mathbf{I}} = (-1)^s$.

PROOF.– By definition, $\overline{\mathbf{I}} = \mathbf{I} : \mathbf{I} = Q_n(\mathbf{I}) = (-1)^s$. For the conjugate of 1, we write, according to the definition:

$$\forall \mathbf{X} \in \Lambda^n \boldsymbol{E}, B_n(\mathbf{I} : 1, \mathbf{X}) = B_n(\mathbf{I}, \mathbf{X} \wedge 1), \qquad [9.33]$$

however, we have set, by definition, that $1 \wedge \mathbf{X} = \mathbf{X} \wedge 1 = \mathbf{X}$. Thus, we have $\mathbf{I} : 1 = \mathbf{I}$.

9.1.2. *Elementary properties of the H-conjugation*

THEOREM 9.2.– The conjugation operator is a *signed* pseudo-isometry between the spaces $(\Lambda^p \boldsymbol{E}, B_p)$ and $(\Lambda^{n-p} \boldsymbol{E}, B_{n-p})$ in the following sense:

$$\forall \mathbf{E}, \mathbf{F} \in \Lambda^p \boldsymbol{E}, \quad B_{n-p}(\overline{\mathbf{E}}, \overline{\mathbf{F}}) = (-1)^s B_p(\mathbf{E}, \mathbf{F}), \qquad [9.34]$$

where s is the signature of the nondegenerate form b over $\boldsymbol{E}$.

PROOF.– We show at first that the conjugations maintain (up to the sign $-(1)^s$) the quadratic form over anisotropic p-blades. Consider thus $\mathbf{x}_1, \cdots, \mathbf{x}_p$ such that $Q_p(\mathbf{x}_1 \wedge \mathbf{x}_p) \neq 0$, we have:

$$B_{n-p}(\mathbf{I} : (\mathbf{x}_1 \wedge \cdots \wedge \mathbf{x}_p), \mathbf{I} : (\mathbf{x}_1 \wedge \cdots \wedge \mathbf{x}_p)). \qquad [9.35]$$

By definition of the contraction product, this equals:

$$B_n(\mathbf{I}, [\mathbf{I} : (\mathbf{x}_1 \wedge \cdots \wedge \mathbf{x}_p)] \wedge (\mathbf{x}_1 \wedge \cdots \wedge \mathbf{x}_p)). \qquad [9.36]$$

Note that, in virtue of formula [3.16], we have the relation:

$$[\mathbf{I}:(\mathbf{x}_1\wedge\cdots\wedge\mathbf{x}_p)]\wedge(\mathbf{x}_1\wedge\cdots\wedge\mathbf{x}_p)=$$
$$(-1)^{p(n-p)}(\mathbf{x}_1\wedge\cdots\wedge\mathbf{x}_p)\wedge\overline{(\mathbf{x}_1\wedge\cdots\wedge\mathbf{x}_p)}.\quad[9.37]$$

By employing the interpretation of the H-conjugate for anisotropic p-blades (see proposition 9.1), we have:

$$(\mathbf{x}_1\wedge\cdots\wedge\mathbf{x}_p)\wedge\overline{(\mathbf{x}_1\wedge\cdots\wedge\mathbf{x}_p)}=(-1)^{p(n-p)}Q_p(\mathbf{x}_1\wedge\cdots\wedge\mathbf{x}_p)\,\mathbf{I}.\quad[9.38]$$

Hence, we deduce:

$$B_n(\mathbf{I},[\mathbf{I}:(\mathbf{x}_1\wedge\cdots\wedge\mathbf{x}_p)]\wedge(\mathbf{x}_1\wedge\cdots\wedge\mathbf{x}_p))=$$
$$Q_p(\mathbf{x}_1\wedge\cdots\wedge\mathbf{x}_p)\,B_n(\mathbf{I},\mathbf{I}),\quad[9.39]$$

in other words,

$$B_n(\mathbf{I},[\mathbf{I}:(\mathbf{x}_1\wedge\cdots\wedge\mathbf{x}_p)]\wedge(\mathbf{x}_1\wedge\cdots\wedge\mathbf{x}_p))=(-1)^s\,Q_p(\mathbf{x}_1\wedge\cdots\wedge\mathbf{x}_p).\quad[9.40]$$

Next, we consider the quadratic bilinear forms associated with the following way. For any non-isotropic p-blades $\mathbf{E},\mathbf{F}$:

$$Q_{n-p}(\mathbf{I}:\mathbf{E}+\mathbf{I}:\mathbf{F})=Q_{n-p}(\mathbf{I}:(\mathbf{E}+\mathbf{F}))=(-1)^s\,Q_p(\mathbf{E}+\mathbf{F}).\quad[9.41]$$

We immediately deduce that:

$$\frac{1}{2}\left[Q_{n-p}\left(\overline{\mathbf{E}}+\overline{\mathbf{F}}\right)-Q_{n-p}\left(\overline{\mathbf{E}}\right)-Q_{n-p}\left(\overline{\mathbf{F}}\right)\right]=$$
$$(-1)^s\,\frac{1}{2}\left[Q_p(\mathbf{E}+\mathbf{F})-Q_p(\mathbf{E})-Q_p(\mathbf{F})\right],\quad[9.42]$$

therefore, the following relation is true for all non-isotropic p-blades:

$$B_{n-p}\left(\overline{\mathbf{E}},\overline{\mathbf{F}}\right)=(-1)^s\,B_p(\mathbf{E},\mathbf{F}).\quad[9.43]$$

Now, since it is possible to find some bases of $\Lambda^p\boldsymbol{E}$ from non-isotropic p-blades, we can, by linearity of the conjugation and by bilinearity of the function B_p, B_{n-p}, extend the relation above to all the p-vectors $\mathbf{E},\mathbf{F}$ of $\Lambda^p\boldsymbol{E}$.

PROPOSITION 9.3.– For all $\mathbf{E} \in \Lambda^p E$, we have the following identity:

$$\overline{\overline{(\mathbf{E})}} = (-1)^s (-1)^{p(n-p)} \mathbf{E}, \qquad [9.44]$$

where s is the signature of the nondegenerate form b over E.

PROOF.– Let $\mathbf{e}_1, \cdots, \mathbf{e}_n$ be a pseudo-orthonormal basis of E, such that $\mathbf{e}_1 \wedge \cdots \wedge \mathbf{e}_n = \mathbf{I}$. Then, we know (see corollary 9.1) that:

$$\overline{\mathbf{e}_1 \wedge \cdots \wedge \mathbf{e}_p} = (-1)^{p(n-p)} q(\mathbf{e}_1) \cdots q(\mathbf{e}_p) \mathbf{e}_{p+1} \wedge \cdots \mathbf{e}_n. \qquad [9.45]$$

Let us now calculate $\overline{\mathbf{e}_{p+1} \wedge \cdots \wedge \mathbf{e}_n}$. We know that:

$$\mathbf{e}_{p+1} \wedge \cdots \wedge \mathbf{e}_n \wedge (\mathbf{e}_1 \wedge \cdots \wedge \mathbf{e}_p) = (-1)^{p(n-p)} \mathbf{I}. \qquad [9.46]$$

And, on the other hand,

$$\mathbf{e}_{p+1} \wedge \cdots \wedge \mathbf{e}_n \wedge \overline{\mathbf{e}_{p+1} \wedge \cdots \wedge \mathbf{e}_n} = (-1)^{p(n-p)} q(\mathbf{e}_{p+1}) \cdots q(\mathbf{e}_n) \mathbf{I}. \qquad [9.47]$$

By identification (which is possible due to the orthogonality with respect to $\mathbf{e}_i, i \in [p+1, n]$), we can then identify:

$$\overline{\mathbf{e}_{p+1} \wedge \cdots \wedge \mathbf{e}_n} = q(\mathbf{e}_{p+1}) \cdots q(\mathbf{e}_n) (\mathbf{e}_1 \wedge \cdots \wedge \mathbf{e}_p). \qquad [9.48]$$

We hence deduce:

$$\overline{\overline{(\mathbf{e}_1 \wedge \cdots \wedge \mathbf{e}_p)}} = (-1)^{p(n-p)} q(\mathbf{e}_1) \cdots q(\mathbf{e}_n) \mathbf{e}_1 \wedge \cdots \wedge \mathbf{e}_p, \qquad [9.49]$$

in virtue of Sylvester's law of inertia (see theorem 6.7), this is equal to:

$$\overline{\overline{(\mathbf{e}_1 \wedge \cdots \wedge \mathbf{e}_p)}} = (-1)^s (-1)^{p(n-p)} \mathbf{e}_1 \wedge \cdots \wedge \mathbf{e}_p, \qquad [9.50]$$

by linearity, this relation can then be extended to all the vectors $\mathbf{E}$ of the space $\Lambda^p E$.

As a conclusion, the following result is illustrated:

PROPOSITION 9.4.– Let $\mathbf{x}_1 \wedge \cdots \wedge \mathbf{x}_p$ be a p-blade such that $Q_p(\mathbf{x}_1 \wedge \cdots \wedge \mathbf{x}_p) \neq 0$. Then, its H-conjugate is still a $(n-p)$-blade.

PROOF.– It suffices to see that anisotropic p-blades are associated with nondegenerate subspaces (see corollary 7.2). Therefore, it is possible to construct, by means of a linear combination of $\mathbf{x}_1, \cdots, \mathbf{x}_p$, a pseudo-orthonormal family $\mathbf{e}_1, \cdots, \mathbf{e}_p$. We have then:

$$\mathbf{x}_1 \wedge \cdots \wedge \mathbf{x}_p = \alpha \mathbf{e}_1 \wedge \cdots \wedge \mathbf{e}_p \Rightarrow \overline{\mathbf{x}_1 \wedge \cdots \wedge \mathbf{x}_p}$$
$$= \alpha (-1)^{p(n-p)} q(\mathbf{e}_1) \cdots q(\mathbf{e}_p) \mathbf{e}_{p+1} \wedge \cdots \wedge \mathbf{e}_n, \qquad [9.51]$$

where the basis $\mathbf{e}_1, \cdots, \mathbf{e}_n$ is a direct pseudo-orthonormal basis. As a result, the conjugate is indeed an $(n-p)$-blade.

9.1.3. *On the confusion regarding the vector cross product in three-dimensional spaces*

In this section, we are going to briefly illustrate the idea of inadequacy of the vector cross product in the context of (Euclidean) geometry. In fact, the greatest issue of the vector cross product consists of the confusion that it creates between the algebras $\Lambda^2\mathbb{R}^3$ and $\Lambda^1\mathbb{R}^3$, in other words, between the spaces $\Lambda^2\mathbb{R}^3$ and $\mathbb{R}^3$ itself. Upon closer examination, the vector cross product has all the features of an exterior product:

1) The product $\mathbf{a} \times \mathbf{b}$ performs indeed a skew-symmetric bilinear operation.

2) All the important geometric interpretations that we have defined for the exterior product are still valid for the vector cross product: its norm (in $\mathbb{R}^3$) represents indeed the area of the parallelogram constructed over the vectors, it is null precisely in the case of collinearity of the vectors, it enables the definition of oriented surfaces, etc.

However, unlike the exterior product, the vector cross product assumes that the result of the operation $a \times b$ *has the same character as its factors*: in other words, the vector cross product acts as an internal operation. Initially, this aspect might seem more simple to handle. Nevertheless, a price to pay consists of the absence of generalization of such a product in other dimensions. In particular, when we are working in $\mathbb{R}^2$, it is impossible to find an internal operation law (a product of two vectors in $\mathbb{R}^2$ which would assume values in $\mathbb{R}^2$) that would maintain the same geometric interpretation as the vector cross product in $\mathbb{R}^3$. In this way, when mechanics is taught, such simple and fundamental notions as the moment of a force or the instantaneous rotation vector do not have two *vector* elements in dimension two equivalent to their definition in dimension three. In order to be able to perform mechanics calculations in dimension two, we are thus always forced to invent a third dimension with more or less "evident" fictitious values to assign to the considered vectors, so that the vector cross product formalism can be re-applied! Actually, the vector cross product in dimension three can be easily understood with the aid of the exterior product and the conjugation, which then provide the possibility of extending the definition to higher dimensions (and to the lower one!). Let us choose a positive orientation for $\mathbb{R}^3$ (equipped with its canonical dot product) and let us set

$\mathbf{I} = \mathbf{e}_1 \wedge \mathbf{e}_2 \wedge \mathbf{e}_3$, with $\mathbf{e}_i$ an orthonormal basis (direct by definition). The choice of the positive orientation enables the definition of the vector cross product between vectors. In that case, it can be verified very easily that we obtain precisely the following theorem:

THEOREM 9.3.– Let $\mathbf{e}_1, \mathbf{e}_2, \mathbf{e}_3$ be an orthonormal basis (supposed direct) of the space $\mathbb{R}^3$, for which we define the vector cross product of two vectors according to the usual formula: if $\mathbf{x} = x_1\mathbf{e}_1 + x_2\mathbf{e}_2 + x_3\mathbf{e}_3$ and if $\mathbf{y} = y_1\mathbf{e}_1 + y_2\mathbf{e}_2 + y_3\mathbf{e}_3$, then:

$$\mathbf{x} \times \mathbf{y} = (x_2y_3 - x_3y_2)\,\mathbf{e}_1 + (x_3y_1 - x_1y_3)\,\mathbf{e}_2 + (x_1y_2 - x_2y_1)\,\mathbf{e}_3, \qquad [9.52]$$

and we have exactly:

$$\mathbf{a} \times \mathbf{b} = \overline{\mathbf{a} \wedge \mathbf{b}}\big|\mathbf{I}, \quad \mathbf{I} = \mathbf{e}_1 \wedge \mathbf{e}_2 \wedge \mathbf{e}_3. \qquad [9.53]$$

PROOF.– From the formal point of view, we have:

$$\mathbf{x} \wedge \mathbf{y} = (x_1y_2 - x_2y_1)\,\mathbf{e}_1 \wedge \mathbf{e}_2 + (x_1y_3 - x_3y_1)\,\mathbf{e}_1 \wedge \mathbf{e}_3 + (x_2y_3 - x_2y_1)\,\mathbf{e}_2 \wedge \mathbf{e}_3, \qquad [9.54]$$

it is possible to apply the (linear) conjugation operator to the summation, obtaining:

$$\overline{\mathbf{x} \wedge \mathbf{y}}\big|\mathbf{I} = (x_1y_2 - x_2y_1)\,\overline{\mathbf{e}_1 \wedge \mathbf{e}_2}\big|\mathbf{I} + (x_1y_3 - x_3y_1)\,\overline{\mathbf{e}_1 \wedge \mathbf{e}_3}\big|\mathbf{I} + (x_2y_3 - x_2y_1)\,\overline{\mathbf{e}_2 \wedge \mathbf{e}_3}\big|\mathbf{I}. \qquad [9.55]$$

We can easily verify that:

$$\overline{\mathbf{e}_1 \wedge \mathbf{e}_2} = \mathbf{e}_3, \quad \overline{\mathbf{e}_1 \wedge \mathbf{e}_3} = -\mathbf{e}_2, \quad \overline{\mathbf{e}_2 \wedge \mathbf{e}_3} = \mathbf{e}_1, \qquad [9.56]$$

and the result follows immediately.

Incidentally, we can thus observe that the "generalization" of the vector cross product in dimension two becomes a scalar. In this (Euclidean) case, the vector cross product and the exterior product exchange the roles of $\mathbb{R}$ and $\Lambda^2\mathbb{R}^2$ by exchanging 1 and $\mathbf{I} = \mathbf{e}_1 \wedge \mathbf{e}_2$ according to:

$$\mathbf{a} \times \mathbf{b} = \overline{\mathbf{a} \wedge \mathbf{b}} = \mathbf{I} : (\mathbf{a} \wedge \mathbf{b}) = \det_{\mathbf{e}_1, \mathbf{e}_2}(\mathbf{a}, \mathbf{b}) \qquad [9.57]$$

for the latter equality, we refer to proposition 9.5 below.

9.1.4. *Demonstration of determinant calculation rules*

First of all, demonstrating the following result will be helpful:

PROPOSITION 9.5.– Let $\mathbf{e}_1, \cdots, \mathbf{e}_n$ be a pseudo-orthonormal direct basis of $(\boldsymbol{E}, b)$, with a signature s. Then, for all family $\mathbf{x}_1, \cdots, \mathbf{x}_n$ of $\boldsymbol{E}$:

$$\det_{\mathbf{e}_1,\cdots,\mathbf{e}_n}(\mathbf{x}_1, \cdots, \mathbf{x}_n) = (-1)^s \mathbf{I} : (\mathbf{x}_1, \cdots, \mathbf{x}_n). \quad [9.58]$$

PROOF.– It is immediate by employing the fact that:

$$\mathbf{x}_1, \cdots, \mathbf{x}_n = \det_{\mathbf{e}_1,\cdots,\mathbf{e}_n}(\mathbf{x}_1, \cdots, \mathbf{x}_n)\, \mathbf{e}_1 \wedge \cdots \mathbf{e}_n = \det_{\mathbf{e}_1,\cdots,\mathbf{e}_n}(\mathbf{x}_1, \cdots, \mathbf{x}_n)\, \mathbf{I}. \quad [9.59]$$

Then, it follows that:

$$\mathbf{I} : (\mathbf{x}_1, \cdots, \mathbf{x}_n) = \det_{\mathbf{e}_1,\cdots,\mathbf{e}_n}(\mathbf{x}_1, \cdots, \mathbf{x}_n)\, \mathbf{I} : \mathbf{I} = (-1)^s \det_{\mathbf{e}_1,\cdots,\mathbf{e}_n}(\mathbf{x}_1, \cdots, \mathbf{x}_n), \quad [9.60]$$

which concludes the demonstration.

In the case of dimension three, we can write simply:

$$\det_{\mathbf{e}_1,\mathbf{e}_2,\mathbf{e}_3}(\mathbf{x}_1, \mathbf{x}_2 \mathbf{x}_3) = \mathbf{I} : (\mathbf{x}_1 \wedge \mathbf{x}_2 \wedge \mathbf{x}_3). \quad [9.61]$$

By employing proposition 8.5 for the contraction product, we obtain:

$$\det_{\mathbf{e}_1,\mathbf{e}_2,\mathbf{e}_3}(\mathbf{x}_1, \mathbf{x}_2, \mathbf{x}_3) = (\mathbf{I} : \mathbf{x}_3) : (\mathbf{x}_1 \wedge \mathbf{x}_2). \quad [9.62]$$

Now, we know that we have $\mathbf{I} : \mathbf{e}_1 = \mathbf{e}_2 \wedge \mathbf{e}_3$, and that $\mathbf{I} : \mathbf{e}_2 = -\mathbf{e}_1 \wedge \mathbf{e}_3$ and also $\mathbf{I} : \mathbf{e}_3 = \mathbf{e}_1 \wedge \mathbf{e}_2$. By writing $\mathbf{x}_i = \sum_{j=1}^{j=3} x_{ji} \mathbf{e}_j$, we obtain immediately:

vectors	$(\mathbf{x}_1 \wedge \mathbf{x}_2)$	$(\mathbf{I} : \mathbf{x}_3)$
$\mathbf{e}_1 \wedge \mathbf{e}_2$	$x_{11}x_{22} - x_{21}x_{12}$	x_{33}
$\mathbf{e}_1 \wedge \mathbf{e}_3$	$x_{11}x_{32} - x_{31}x_{12}$	$-x_{32}$
$\mathbf{e}_2 \wedge \mathbf{e}_3$	$x_{21}x_{32} - x_{31}x_{22}$	x_{31}

[9.63]

the contraction product being now equal to the dot product over p-vectors with the same dimension, and since the basis of $\mathbf{e}_i \wedge \mathbf{e}_j$ is orthonormal for the dot product over $\Lambda^2\mathbb{R}^3$, we easily deduce thereof:

$$(\mathbf{I} : \mathbf{x}_3) : (\mathbf{x}_1 \wedge \mathbf{x}_2) = x_{33}(x_{11}x_{22} - x_{21}x_{12}) - x_{32}(x_{11}x_{32} - x_{31}x_{12}) + x_{31}(x_{21}x_{32} - x_{31}x_{22}), \quad [9.64]$$

which corresponds precisely to the determinant. In the case of dimension 4, the calculation principle is exactly the same:

$$\det_{\mathbf{e}_1,\mathbf{e}_2,\mathbf{e}_3,\mathbf{e}_4}(\mathbf{x}_1,\mathbf{x}_2\mathbf{x}_3,\mathbf{x}_4) = \mathbf{I} : (\mathbf{x}_1 \wedge \mathbf{x}_2 \wedge \mathbf{x}_3 \wedge \mathbf{x}_4). \qquad [9.65]$$

Thanks to the formula of the contraction product (see proposition 8.5), we hence deduce:

$$\begin{aligned}\det_{\mathbf{e}_1,\mathbf{e}_2,\mathbf{e}_3,\mathbf{e}_4}(\mathbf{x}_1,\mathbf{x}_2,\mathbf{x}_3,\mathbf{x}_4) &= ((\mathbf{I} : \mathbf{x}_4) : \mathbf{x}_3) : (\mathbf{x}_1 \wedge \mathbf{x}_2) \\ &= (\mathbf{I} : (\mathbf{x}_3 \wedge \mathbf{x}_4)) : (\mathbf{x}_1 \wedge \mathbf{x}_2). \qquad [9.66]\end{aligned}$$

Therefore, we know how to calculate the exterior product, as well as the dot product. It remains to examine the conjugation. For this purpose, it suffices to observe:

$$\begin{array}{cccccc}\overline{\mathbf{e}_1 \wedge \mathbf{e}_2} & \overline{\mathbf{e}_1 \wedge \mathbf{e}_3} & \overline{\mathbf{e}_1 \wedge \mathbf{e}_4} & \overline{\mathbf{e}_2 \wedge \mathbf{e}_3} & \overline{\mathbf{e}_2 \wedge \mathbf{e}_4} & \overline{\mathbf{e}_3 \wedge \mathbf{e}_4} \\ \mathbf{e}_3 \wedge \mathbf{e}_4 & -\mathbf{e}_2 \wedge \mathbf{e}_4 & \mathbf{e}_2 \wedge \mathbf{e}_3 & \mathbf{e}_1 \wedge \mathbf{e}_4 & -\mathbf{e}_1 \wedge \mathbf{e}_3 & \mathbf{e}_1 \wedge \mathbf{e}_2\end{array}. \qquad [9.67]$$

We thus see that by the inverting lexicographic order, up to a sign, we obtain a conjugation. This fact was already true in the case of dimension three, and for this reason, the calculation rules are eventually analogous. As a result, the column representation, which was performed in section 5.2.5, can be actually applied.

9.2. Regressive product

9.2.1. *Definition*

DEFINITION 9.2.– Let $\mathbf{E} \in \Lambda^p E$ and $\mathbf{F} \in \Lambda^q E$. We define by the symbol "$(\cdot) \vee (\cdot)$":

$$\overline{\mathbf{E} \vee \mathbf{F}} := \overline{\mathbf{E}} \wedge \overline{\mathbf{F}} \qquad [9.68]$$

as being, by definition, the regressive product (also called "meet" or "anti-wedge" product) of $\mathbf{E}$ by $\mathbf{F}$.

REMARK 9.2.– Before proceeding further, let us point out:

1) The definition is assured to the extent that the duality operator is an isomorphism between $\Lambda^k E$ and $\Lambda^{n-k} E$: we can thus define an element by defining its image via the usage of this isomorphism.

2) While the exterior product is a progressive product, in the sense that the degree of the product is most often equal to the sum of the degrees, the regressive product is such that its degree is lower than the degree of its factors. According to the definition that we have introduced, the degree of $\mathbf{E} \vee \mathbf{F}$ is zero if $p+q<n$, and it equals $p+q-n$ if $n \leq p+q \leq 2n$, in which case the degree of the product is indeed lower than the sum of the degrees. ■

It is possible to provide immediately another formulation of the exterior product:

PROPOSITION 9.6.– We have the following equivalent formulation:

$$\forall \mathbf{E} \in \Lambda^p E, \forall \mathbf{F} \in \Lambda^q E, \ \ \mathbf{E} \vee \mathbf{F} = (-1)^{(p+q)(p+q-n)+s}\, \overline{\overline{\mathbf{E}} \wedge \overline{\mathbf{F}}}. \qquad [9.69]$$

PROOF.– If $p+q-n<0$, then $\overline{\mathbf{E}} \wedge \overline{\mathbf{F}} = \mathbf{0}$ and as a result, its conjugate is also null: the formula is true. If now $p+q-n \geq 0$, then we know that $\mathbf{E} \vee \mathbf{F}$ is an element of Λ^{p+q-n}. We can thus write:

$$\overline{\overline{\mathbf{E} \vee \mathbf{F}}} = \overline{\overline{\mathbf{E}} \wedge \overline{\mathbf{F}}}. \qquad [9.70]$$

Now, by applying the conjugation formula (proposition 9.3), we obtain:

$$\overline{\overline{\mathbf{E} \vee \mathbf{F}}} = (-1)^{(p+q-n)(n-(p+q-n))+s}\, \mathbf{E} \vee \mathbf{F}, \qquad [9.71]$$

and we have indeed the formula of the proposition.

The following corollary is deduced from the above expression:

COROLLARY 9.2.– For every pair $\mathbf{E} \in \Lambda^p E$ and $\mathbf{F} \in \Lambda^q E$, we have:

$$\overline{\mathbf{E}} \vee \overline{\mathbf{F}} = (-1)^s\, \overline{\mathbf{E} \wedge \mathbf{F}}. \qquad [9.72]$$

PROOF.– According to the previous formula, we have:

$$\overline{\mathbf{E}} \vee \overline{\mathbf{F}} = (-1)^{(p+q)(p+q-n)+s}\, \overline{\overline{\overline{\mathbf{E}}} \wedge \overline{\overline{\mathbf{F}}}}. \qquad [9.73]$$

We know that:

$$\overline{\overline{\mathbf{E}}} = (-1)^{p(n-p)+s}\, \mathbf{E}, \ \ \overline{\overline{\mathbf{F}}} = (-1)^{q(n-q)+s}\, \mathbf{F}. \qquad [9.74]$$

Thus, we actually have:

$$\overline{\mathbf{E}} \vee \overline{\mathbf{F}} = (-1)^{\alpha}\, \overline{\overline{\mathbf{E}} \wedge \overline{\mathbf{F}}}, \qquad [9.75]$$

and it remains to determine the parity of α. Indeed, we have:

$$\begin{aligned}\alpha &= (p+q)(p+q-n) + p(n-p) + q(n-q) + 3s \\ &= (p+q)^2 - p^2 - q^2 + 3s = 2(pq+s) + s,\end{aligned} \qquad [9.76]$$

hence the stated result.

After the definition, it is equally possible to show that:

PROPOSITION 9.7.– For all $\mathbf{E}$ of $\Lambda^p \boldsymbol{E}$ and all $\mathbf{F}$ in $\Lambda^q \boldsymbol{E}$, we have:

$$\mathbf{E} \vee \mathbf{F} = (-1)^{(n-p)(n-q)}\, \mathbf{F} \vee \mathbf{E}. \qquad [9.77]$$

PROOF.– Actually, we have:

$$\overline{\mathbf{E} \vee \mathbf{F}} = \overline{\mathbf{E}} \wedge \overline{\mathbf{F}} = (-1)^{(n-p)(n-q)}\, \overline{\mathbf{F}} \wedge \overline{\mathbf{E}} = (-1)^{(n-p)(n-q)}\, \overline{\mathbf{F} \vee \mathbf{E}}, \qquad [9.78]$$

and, since the H-conjugation is invertible, we hence deduce our result.

Let us demonstrate straight away a very important property of the regressive product:

PROPOSITION 9.8.– The regressive product law is associative: for any triplet $\mathbf{E}, \mathbf{F}, \mathbf{G}$, respectively p, q, r-vectors over $\boldsymbol{E}$ (with a dimension n), we have:

$$(\mathbf{E} \vee \mathbf{F}) \vee \mathbf{G} = \mathbf{E} \vee (\mathbf{F} \vee \mathbf{G}). \qquad [9.79]$$

PROOF.– The idea here consists of expressing the regressive products exclusively with exterior products: thus, since we know that the exterior product is associative, we shall know that the regressive product is itself associative. According to the definition of the regressive product, we have:

$$\overline{\mathbf{E} \vee \mathbf{F}} = \overline{\mathbf{E}} \wedge \overline{\mathbf{F}}. \qquad [9.80]$$

Now, thanks to proposition 9.6, we also know that:

$$\begin{aligned}(\mathbf{E} \vee \mathbf{F}) \vee \mathbf{G} &= (-1)^{\alpha(p,q,r)+s}\, \overline{\left(\overline{\mathbf{E} \vee \mathbf{F}} \wedge \overline{\mathbf{G}}\right)} \\ &= (-1)^{s}(-1)^{\alpha(p,q,r)}\, \overline{\left(\overline{\mathbf{E}} \wedge \overline{\mathbf{F}}\right) \wedge \overline{\mathbf{G}}}.\end{aligned} \qquad [9.81]$$

The associativity of the left-hand side is ensured if the associativity of the right-hand side is shown. The exterior product being associative, we already have the first part of the answer. In order to provide a definitive answer, we need to study the function $(-1)^{\alpha(p,q,r)}$ and show that it is associative as well. By referring to proposition 9.6, we have:

$$\begin{aligned}\alpha(p,q,r) &= [p+q-n+r][(p+q-n)+r-n] \\ &= (p+q+r-2n)(p+q+r-n).\end{aligned} \quad [9.82]$$

Since we are looking for the value of α in $\mathbb{Z}/2\mathbb{Z}$, we can write:

$$\alpha(p,q,r) \equiv (p+q+r)(p+q+r-n), \quad [9.83]$$

which is indeed associative with respect to p, q, r. We obtain thus the associativity of the regressive product.

9.2.2. *Some calculation formulas*

Let us begin with two interesting formulas:

PROPOSITION 9.9.– Let $\mathbf{A}, \mathbf{B}$ be two p-vectors of $\Lambda^p \boldsymbol{E}$, B_p. Then, we have:

$$\overline{\mathbf{A}} \wedge \mathbf{B} = B_p(\mathbf{A},\mathbf{B})\,\mathbf{I} \quad [9.84]$$

$$\overline{\mathbf{A}} \vee \mathbf{B} = (-1)^s B_p(\mathbf{A},\mathbf{B}) \quad [9.85]$$

PROOF.– Note, in fact, that:

$$\overline{\overline{\mathbf{A}} \wedge \mathbf{B}} = \mathbf{I} : \left(\overline{\mathbf{A}} \wedge \mathbf{B}\right) = B_n\left(\mathbf{I}, \overline{\mathbf{A}} \wedge \mathbf{B}\right). \quad [9.86]$$

By employing the definition of the contraction product (see definition 8.1), we obtain:

$$B_n\left(\mathbf{I}, \overline{\mathbf{A}} \wedge \mathbf{B}\right) = B_{n-p}\left(\mathbf{I} : \mathbf{B}, \overline{\mathbf{A}}\right) = B_{n-p}\left(\overline{\mathbf{A}}, \overline{\mathbf{B}}\right) = (-1)^s B_p(\mathbf{A},\mathbf{B}), \quad [9.87]$$

now, in virtue of proposition 9.2, we have $\overline{1} = \mathbf{I}$ and $\overline{\mathbf{I}} = (-1)^s$. As a result:

$$\overline{\overline{\overline{\mathbf{A}} \wedge \mathbf{B}}} = (-1)^{n(n-n)+s}\, \overline{\mathbf{A}} \wedge \mathbf{B} = \overline{(-1)^s B_p(\mathbf{A},\mathbf{B})} = (-1)^s B_p(\mathbf{A},\mathbf{B})\,\mathbf{I}, \quad [9.88]$$

hence finally:

$$\overline{\mathbf{A}} \wedge \mathbf{B} = B_p(\mathbf{A}, \mathbf{B})\, \mathbf{I}. \qquad [9.89]$$

As for the second formula, initially we write that:

$$\overline{\mathbf{A}} \vee \mathbf{B} = (-1)^{p(n-p)+s}\, \overline{\mathbf{A}} \vee \overline{\overline{\mathbf{B}}} := (-1)^{p(n-p)+s} (-1)^s\, \overline{\mathbf{A} \wedge \overline{\mathbf{B}}}, \qquad [9.90]$$

the latter equality being obtained via corollary 9.2. However, it is also possible to write that:

$$\mathbf{A} \wedge \overline{\mathbf{B}} = (-1)^{p(n-p)}\, \overline{\mathbf{B}} \wedge \mathbf{A} = (-1)^{p(n-p)}\, B_p(\mathbf{A}, \mathbf{B})\, \mathbf{I}, \qquad [9.91]$$

hence, we deduce:

$$\overline{\mathbf{A}} \vee \mathbf{B} = \overline{B_p(\mathbf{A}, \mathbf{B})\, \mathbf{I}} = (-1)^s\, B_p(\mathbf{A}, \mathbf{B}) \qquad [9.92]$$

with the last equality obtained by means of $\overline{\mathbf{I}} = \mathbf{I} : \mathbf{I} = (-1)^s$.

A straightforward corollary generalizes proposition 9.1, which until now was demonstrated only in the case of non-isotropic p-blades:

COROLLARY 9.3.– Let $\mathbf{E} \in \Lambda^p E$. Then, $\overline{\mathbf{E}} \wedge \mathbf{E} = Q_p(\mathbf{E})\, \mathbf{I}$.

Let us now address the formula of the double product:

PROPOSITION 9.10.– Let $\mathbf{x}, \mathbf{y}, \mathbf{z}$ be three vectors of (E, b), with a dimension n and a signature s. Then, we have:

$$(\mathbf{x} \wedge \mathbf{y}) \vee \overline{\mathbf{z}} = (-1)^{n+s} \left[b(\mathbf{z}, \mathbf{y})\, \mathbf{x} - b(\mathbf{z}, \mathbf{x})\, \mathbf{y} \right]. \qquad [9.93]$$

PROOF.– The degree of $(\mathbf{x} \wedge \mathbf{y}) \vee \overline{\mathbf{z}}$ is $2 + (n-1) - n = 1$. Let us remind that:

$$\overline{(\mathbf{x} \wedge \mathbf{y}) \vee \overline{\mathbf{z}}} = \overline{\mathbf{x} \wedge \mathbf{y}} \wedge \overline{\overline{\mathbf{z}}} = (-1)^{n-1+s}\, \overline{\mathbf{x} \wedge \mathbf{y}} \wedge \mathbf{z}, \qquad [9.94]$$

hence, we deduce in fact:

$$(\mathbf{x} \wedge \mathbf{y}) \vee \overline{\mathbf{z}} = (-1)^{n-1+s}\, \overline{\overline{(\mathbf{x} \wedge \mathbf{y}) \vee \overline{\mathbf{z}}}} = \overline{\overline{\mathbf{x} \wedge \mathbf{y}} \wedge \mathbf{z}}. \qquad [9.95]$$

By considering the pseudo-dot product with a vector $\mathbf{t}$, we obtain:

$$((\mathbf{x} \wedge \mathbf{y}) \vee \overline{\mathbf{z}}) : \mathbf{t} = [\mathbf{I} : (\overline{\mathbf{x} \wedge \mathbf{y}} \wedge \mathbf{z})] : \mathbf{t} = ([\mathbf{I} : \mathbf{z}] : (\overline{\mathbf{x} \wedge \mathbf{y}})) : \mathbf{t}, \qquad [9.96]$$

which can be rewritten in the form:

$$(\mathbf{I} : \mathbf{z}) : (\mathbf{t} \wedge (\overline{\mathbf{x} \wedge \mathbf{y}})), \qquad [9.97]$$

now, we can verify that $\mathbf{t} \wedge (\overline{\mathbf{x} \wedge \mathbf{y}}) = (-1)^{n-2} \overline{\mathbf{x} \wedge \mathbf{y}} \wedge \mathbf{t}$, hence we deduce:

$$\begin{aligned}(\mathbf{I} : \mathbf{z}) : (\mathbf{t} \wedge (\overline{\mathbf{x} \wedge \mathbf{y}})) &= (-1)^n [(\mathbf{I} : \mathbf{z}) : \mathbf{t}] : \overline{\mathbf{x} \wedge \mathbf{y}} \\ &= (-1)^n [\mathbf{I} : \mathbf{t} \wedge \mathbf{z}] : \overline{\mathbf{x} \wedge \mathbf{y}} = (-1)^n \left(\overline{\mathbf{t} \wedge \mathbf{z}} : \overline{\mathbf{x} \wedge \mathbf{y}}\right). \qquad [9.98]\end{aligned}$$

By employing the fact that the H-conjugation is a signed pseudo-isometry (see theorem 9.2), it entails immediately that:

$$\overline{\mathbf{t} \wedge \mathbf{z}} : \overline{\mathbf{x} \wedge \mathbf{y}} = (-1)^s \, \mathbf{t} \wedge \mathbf{z} : \mathbf{x} \wedge \mathbf{y}. \qquad [9.99]$$

It is possible to expand the pseudo-dot product in order to obtain:

$$\mathbf{t} \wedge \mathbf{z} : \mathbf{x} \wedge \mathbf{y} = b(\mathbf{t}, \mathbf{x})\, b(\mathbf{z}, \mathbf{y}) - b(\mathbf{t}, \mathbf{y})\, b(\mathbf{z}, \mathbf{x}). \qquad [9.100]$$

Hence, we deduce that for all vector $\mathbf{t}$:

$$b((\mathbf{x} \wedge \mathbf{y}) \vee \overline{\mathbf{z}}, \mathbf{t}) = (-1)^{n+s}\, b(b(\mathbf{z}, \mathbf{y})\, \mathbf{x} - b(\mathbf{z}, \mathbf{x})\, \mathbf{y}, \mathbf{t}). \qquad [9.101]$$

Since the form b is nondegenerate, we deduce thereof that:

$$(\mathbf{x} \wedge \mathbf{y}) \vee \overline{\mathbf{z}} = (-1)^{n+s} [b(\mathbf{z}, \mathbf{y})\, \mathbf{x} - b(\mathbf{z}, \mathbf{x})\, \mathbf{y}]. \qquad [9.102]$$

Note the following aspect:

REMARK 9.3.– The formula of double product is equivalent, in dimension three, to the formula of double vector cross product (in the Euclidean case). Thus, we remind that:

$$(\mathbf{x} \times \mathbf{y}) \times \mathbf{z} = (\mathbf{x} \cdot \mathbf{z})\, \mathbf{y} - (\mathbf{y} \cdot \mathbf{z})\, \mathbf{x}, \qquad [9.103]$$

which is obtained precisely from the previous formula by setting $n = 3$, $s = 0$ (and thus $(-1)^{3+0} = -1$). In order to observe that it suffices to rewrite the vector cross product starting from the conjugation and the exterior product:

$$(\mathbf{x} \times \mathbf{y}) \times \mathbf{z} = \overline{\overline{\mathbf{x} \wedge \mathbf{y}} \wedge \mathbf{z}} = (\mathbf{x} \wedge \mathbf{y}) \vee \overline{\mathbf{z}}, \qquad [9.104]$$

which is exactly formula [9.94] with $n = 3$, $s = 0$. ■

10

Endomorphisms of Exterior Algebras

Exterior algebra is an important tool for studying endomorphisms over $\boldsymbol{E}$. In particular, in the same way as vectors of $\boldsymbol{E}$ are employed to construct vectors of $\Lambda^p\boldsymbol{E}$, we address the following question:

QUESTION 10.1.– To what extent can endomorphisms of $\mathcal{L}(\boldsymbol{E})$ be used to construct endomorphisms of $\mathcal{L}(\Lambda^p\boldsymbol{E})$? In what way do linear maps of $\mathcal{L}(\Lambda^p\boldsymbol{E})$ allow us to understand the maps of $\mathcal{L}(\boldsymbol{E})$?

Behind these spontaneously formulated questions, we shall discover a formalism of invariants and the Laplace inversion formula for endomorphisms.

10.1. Constructible endomorphisms

10.1.1. *Construction of an endomorphism over* $\Lambda^p\boldsymbol{E}$

It is possible to construct endomorphisms over $\Lambda^p\boldsymbol{E}$ starting from endomorphisms of $\boldsymbol{E}$ in a rather general way. In fact, we shall see that there are basically two methods for achieving this: one via a symmetric construction; the other via a skew-symmetric construction. The latter actually enables the definition of an exterior product over linear operators. In a sense, this construction is quite natural. The first construction, on the other hand, proves to be both a surprising and useful tool for the study of endomorphisms over $\boldsymbol{E}$:

THEOREM 10.1.– Let $f_1, \cdots, f_p$ be a set of p endomorphisms over $\boldsymbol{E}$. Then, the map defined by

$$\forall \mathbf{x}_1, \cdots, \mathbf{x}_p \in \boldsymbol{E}^n, \quad \mathcal{S}(\mathbf{x}_1, \cdots, \mathbf{x}_p) = \frac{1}{p!} \sum_{\alpha \in S_p} f_{\alpha(1)}(\mathbf{x}_1) \wedge \cdots \wedge f_{\alpha(p)}(\mathbf{x}_p) \quad [10.1]$$

is an alternating p-linear map over $\boldsymbol{E}^p$ with values in $\Lambda^p \boldsymbol{E}$.

PROOF.– The multi-linear aspect is evident since, for a given permutation $\alpha \in S_p$, the map

$$(\mathbf{x}_1, \cdots, \mathbf{x}_p) \mapsto f_{\alpha(1)}(\mathbf{x}_1) \wedge \cdots \wedge f_{\alpha(p)}(\mathbf{x}_p) \quad [10.2]$$

is obviously multi-linear (although not alternating). By performing a summation over the permutations, we still have a multi-linear map. Furthermore, the summation over the permutations reintroduces the skew-symmetric aspect of the map: let us choose $\sigma \in S_p$; then, it is possible to calculate:

$$\mathcal{S}\left(\mathbf{x}_{\sigma(1)}, \cdots, \mathbf{x}_{\sigma(p)}\right) = \frac{1}{p!} \sum_{\alpha \in S_p} f_{\alpha(1)}\left(\mathbf{x}_{\sigma(1)}\right) \wedge \cdots \wedge f_{\alpha(p)}\left(\mathbf{x}_{\sigma(p)}\right). \quad [10.3]$$

If the elements $\mathbf{x}_{\sigma(i)}$ are arranged in increasing order, it can be easily verified that:

$$\mathcal{S}\left(\mathbf{x}_{\sigma(1)}, \cdots, \mathbf{x}_{\sigma(p)}\right) = \frac{1}{p!} \sum_{\alpha \in S_p} \varepsilon(\sigma) f_{\alpha \circ \sigma^{-1}(1)}(\mathbf{x}_1) \wedge \cdots \wedge f_{\alpha \circ \sigma^{-1}(p)}(\mathbf{x}_p), \quad [10.4]$$

where $\varepsilon(\sigma)$ is the signature of the permutation σ. Therefore, we can perform a "change of variables" $\alpha \leftrightarrow \alpha \circ \sigma^{-1} = \beta$ within the summation, by obtaining straightaway

$$\begin{aligned} \mathcal{S}\left(\mathbf{x}_{\sigma(1)}, \cdots, \mathbf{x}_{\sigma(p)}\right) &= \varepsilon(\sigma) \frac{1}{p!} \sum_{\beta \in S_p} f_{\beta(1)}(\mathbf{x}_1) \wedge \cdots \wedge f_{\beta(p)}(\mathbf{x}_p) \\ &= \varepsilon(\sigma) \mathcal{S}(\mathbf{x}_1, \cdots, \mathbf{x}_p). \end{aligned} \quad [10.5]$$

Hence, we deduce the following definition:

DEFINITION 10.1.– Let $f_1, \cdots, f_p$ be a sequence of p endomorphisms of $\boldsymbol{E}$. We call a (symmetric) bracket of these endomorphisms, denoted by $[f_1 \cdots f_p]$, the unique endomorphism of $\Lambda^p \boldsymbol{E}$ such that:

$$\forall (\mathbf{x}_1, \cdots, \mathbf{x}_p) \in \boldsymbol{E}, [f_1 \cdots f_n] (\mathbf{x}_1 \wedge \cdots \wedge \mathbf{x}_p)$$
$$= \frac{1}{p!} \sum_{\alpha \in S_p} f_{\alpha(1)} (\mathbf{x}_1) \wedge \cdots \wedge f_{\alpha(p)} (\mathbf{x}_p). \quad [10.6]$$

PROOF.– Since the map

$$\mathcal{S} (\mathbf{x}_1, \cdots, \mathbf{x}_p) \mapsto \frac{1}{p!} \sum_{\alpha \in S_p} f_{\alpha(1)} (\mathbf{x}_1) \wedge \cdots \wedge f_{\alpha(p)} (\mathbf{x}_p) \in \Lambda^p \boldsymbol{E} \quad [10.7]$$

is an alternating multi-linear map, in virtue of the fundamental theorem of factorization of the algebra $\Lambda^p \boldsymbol{E}$ (see theorem 2.1), there exists a unique linear map of $\Lambda^p \boldsymbol{E}$ with values in $\Lambda^p \boldsymbol{E}$ that is coincident with $\mathcal{F}$ over the p-blades. This map is the one in the definition.

The following properties entail immediately:

PROPOSITION 10.1.– The following statements are true:

1) The bracket symbol is commutative with respect to its arguments:

$$\forall \sigma \in S_p, \ \forall f_1, \cdots, f_p \ \left[f_{\sigma(1)} \cdots f_{\sigma(p)}\right] = [f_1 \cdots f_p]. \quad [10.8]$$

2) It always holds that:

$$[I_d \cdots I_d] = I_d \ \textit{over} \ \Lambda^p \boldsymbol{E}. \quad [10.9]$$

3) The bracket symbol satisfies:

$$\forall f_1, \cdots, f_p, \ [f_1 \cdots f_p]^* = \left[f_1^* \cdots f_p^*\right]. \quad [10.10]$$

PROOF.– Let us demonstrate these properties:

1) For any family $\mathbf{x}_1, \cdots, \mathbf{x}_p$, we have:

$$\left[f_{\sigma(1)} \cdots f_{\sigma(p)}\right] = \frac{1}{p!} \sum_{\alpha \in S_p} f_{\alpha(\sigma(1))} (\mathbf{x}_1) \wedge \cdots \wedge f_{\alpha(\sigma(p))} (\mathbf{x}_p), \quad [10.11]$$

then, it suffices to perform a change of variable $\beta \leftrightarrow \alpha \circ \sigma$ in the summation over the permutations and we obtain:

$$\left[f_{\sigma(1)} \cdots f_{\sigma(p)}\right] = \frac{1}{p!} \sum_{\beta \in S_p} f_{\beta(1)}\left(\mathbf{x}_1\right) \wedge \cdots \wedge f_{\beta(p)}\left(\mathbf{x}_p\right), \tag{10.12}$$

hence the symmetry.

2) Immediate.

3) Let us consider a p-blade $\mathbf{y}_1 \wedge \cdots \wedge \mathbf{y}_p$. We calculate then:

$$B_p\left(\left[f_1 \cdots f_p\right]\left(\mathbf{x}_1 \wedge \cdots \wedge \mathbf{x}_p\right), \mathbf{y}_1 \wedge \cdots \wedge \mathbf{y}_p\right), \tag{10.13}$$

this is equal to:

$$\frac{1}{p!} \sum_{\alpha \in S_p} B_p\left(f_{\alpha(1)}\left(\mathbf{x}_1\right) \wedge \cdots \wedge f_{\alpha(1)}\left(\mathbf{x}_1\right), \mathbf{y}_1 \wedge \cdots \wedge \mathbf{y}_p\right), \tag{10.14}$$

which can be calculated via the formula from theorem 7.1, according to:

$$\frac{1}{p!} \sum_{\alpha \in S_p} \det\left(b\left(f_{\alpha(i)}\left(\mathbf{x}_i\right), \mathbf{y}_j\right)\right). \tag{10.15}$$

Thus, we obtain, by definition of pseudo-adjoint, that this is equal to:

$$\frac{1}{p!} \sum_{\alpha \in S_p} \det\left(b\left(\mathbf{x}_i, f_{\alpha(i)}^*\left(\mathbf{y}_j\right)\right)\right)$$
$$= \frac{1}{p!} \sum_{\alpha \in S_p} B_p\left(\mathbf{x}_1 \wedge \cdots \mathbf{x}_p, f_{\alpha(1)}^*\left(\mathbf{y}_1\right) \wedge \cdots \wedge f_{\alpha(n)}^*\left(\mathbf{y}_p\right)\right). \tag{10.16}$$

This, therefore, is equal to:

$$B_p\left[\left(\mathbf{x}_1 \wedge \cdots \wedge \mathbf{x}_p\right), \left[f_1^* \cdots f_p^*\right]\left(\mathbf{y}_1 \wedge \cdots \wedge \mathbf{y}_p\right)\right]. \tag{10.17}$$

Hence, we deduce by identification:

$$\left[f_1 \cdots f_p\right]^* = \left[f_1^* \cdots f_p^*\right]. \tag{10.18}$$

In this way, we reach the following definition:

DEFINITION 10.2.– If within a bracket (of symmetry) we have $u_1 = \cdots = u_p = u$, we then set: $U = [u \cdots u] := [u]_p$. In that case:

$$\forall \mathbf{x}_1, \cdots, \mathbf{x}_p, \ [u]_p (\mathbf{x}_1 \wedge \cdots \wedge \mathbf{x}_p) = u(\mathbf{x}_1) \wedge \cdots \wedge u(\mathbf{x}_p). \qquad [10.19]$$

Let us present some properties of the operator $[\cdot]_p$ when it is applied to other operators:

PROPOSITION 10.2.– The following statements are true:

1) For all f, g endomorphisms of $\boldsymbol{E}$, we have the property: $[f \circ g]_p = [f]_p \circ [g]_p$.

2) The endomorphism $[f]_p$ is invertible over $\Lambda^p \boldsymbol{E}$ if and only if f is invertible. In that case: $[f]_p^{-1} = [f^{-1}]_p$.

3) For all endomorphisms, we have: $[f]_p^* = [f^*]_p$.

PROOF.– Let us demonstrate these properties:

1) We can write, for a generic family:

$$\begin{aligned}[f \circ g]_p (\mathbf{x}_1 \wedge \cdots \wedge \mathbf{x}_p) &= (f \circ g)(\mathbf{x}_1) \wedge \cdots \wedge (f \circ g)(\mathbf{x}_p) \\ &= f(g(\mathbf{x}_1)) \wedge \cdots \wedge f(g(\mathbf{x}_p)),\end{aligned} \qquad [10.20]$$

which can be easily rewritten as:

$$[f]_p (g(\mathbf{x}_1) \wedge \cdots \wedge g(\mathbf{x}_p)) = [f]_p \circ [g]_p (\mathbf{x}_1 \wedge \cdots \wedge \mathbf{x}_p), \qquad [10.21]$$

hence the result.

2) According to the previous property, we immediately see that $[f]_p \circ [f^{-1}]_p = [I_d]_p = I_d$; therefore, if f is invertible, then $[f]_p$ is invertible and its inverse equals $[f^{-1}]_p$. Suppose now that $[f]_p$ is invertible. Let us suppose that f is not invertible, then there exists $\mathbf{x} \neq \mathbf{0}$ such that $f(\mathbf{x}) = \mathbf{0}$. We complete $\mathbf{x}$ with $p-1$ vectors to obtain a linearly independent family. Then, we have:

$$\begin{aligned}&\mathbf{x} \wedge \mathbf{y}_1 \wedge \cdots \wedge \mathbf{y}_{p-1} \neq \mathbf{0} \ \textit{and} \\ &\qquad [f]_p (\mathbf{x} \wedge \mathbf{y}_1 \wedge \cdots \wedge \mathbf{y}_{p-1}) = f(\mathbf{x}) \wedge \cdots \wedge f(\mathbf{y}_n) = \mathbf{0},\end{aligned} \qquad [10.22]$$

which contradicts the fact that $[f]_p$ should be invertible.

3) It suffices to apply formula [10.10] with $f_1 = \cdots = f_p = f$.

Nevertheless, it is quite surprising to work with symmetry (due to the bracket symbol), whereas we have been working with skew symmetry from the very beginning. Well, never mind: instead of making formula [10.1] symmetric, we can simply make it skew-symmetric:

THEOREM 10.2.– Let $f_1, \cdots, f_p$ be a family of p endomorphisms of $\boldsymbol{E}$. Then, the map defined by

$$\forall \mathbf{x}_1, \cdots, \mathbf{x}_p \in \boldsymbol{E}^n, \ \mathcal{A}(\mathbf{x}_1, \cdots, \mathbf{x}_p)$$
$$:= \frac{1}{p!} \sum_{\alpha \in S_p} \varepsilon(\alpha) f_{\alpha(1)}(\mathbf{x}_1) \wedge \cdots \wedge f_{\alpha(p)}(\mathbf{x}_p) \qquad [10.23]$$

is an alternating multi-linear map over $\boldsymbol{E}^p$.

PROOF.– The demonstration is identical to the one illustrated for the symmetric bracket, except that in the conclusion we use the fact that for every pair α, σ of permutations, we have:

$$\varepsilon\left(\alpha \circ \sigma^{-1}\right) = \varepsilon(\alpha)\,\varepsilon\left(\sigma^{-1}\right) = \varepsilon(\alpha)\,\varepsilon(\sigma), \qquad [10.24]$$

and then we can conclude without any effort.

Then, we instantly obtain the following definition:

DEFINITION 10.3.– Let $f_1, \cdots, f_p$ be a sequence of p endomorphisms of $\boldsymbol{E}$. We call exterior product of these endomorphisms, denoted by $f_1 \wedge \cdots \wedge f_p$, the linear map over $\Lambda^p \boldsymbol{E}$, defined by:

$$\forall (\mathbf{x}_1, \cdots, \mathbf{x}_p) \in \boldsymbol{E}^p, \ (f_1 \wedge \cdots \wedge f_p)(\mathbf{x}_1 \wedge \cdots \wedge \mathbf{x}_p) :=$$
$$\frac{1}{p!} \sum_{\alpha \in S_p} \varepsilon(\alpha) f_{\alpha(1)}(\mathbf{x}_1) \wedge \cdots \wedge f_{\alpha(p)}(\mathbf{x}_p). \qquad [10.25]$$

PROOF.– As usual, we apply the fundamental theorem of exterior algebra's construction.

Furthermore, by using the same reasoning as in the case of the bracket symbol, we can show the following property:

PROPOSITION 10.3.– The exterior product symbol is a skew-symmetric symbol:

$$\forall \sigma \in S_p, \ \forall u_1, \cdots, u_p \in \mathcal{L}(\boldsymbol{E}), \ u_{\sigma(1)} \wedge \cdots \wedge u_{\sigma(p)} = \varepsilon(\sigma) \, u_1 \wedge \cdots \wedge u_p. \quad [10.26]$$

PROOF.– The demonstration is equivalent to the previous demonstrations.

10.1.2. *Invariants of endomorphism families*

Now, we are going to employ the symbols presented above to derive some algebraic quantities that are important in the context of endomorphisms:

DEFINITION 10.4.– Let $\boldsymbol{E}$ be a linear space with a dimension n and let $u_1, \cdots, u_n$ be a family of n endomorphisms over $\boldsymbol{E}$. Then, there exists a unique real, denoted by $I(u_1 \cdots u_n)$, such that for all family $\mathbf{x}_1, \cdots, \mathbf{x}_n$, we have:

$$[u_1 \cdots u_n](\mathbf{x}_1 \wedge \cdots \wedge \mathbf{x}_n) = I(u_1 \cdots u_n) \, \mathbf{x}_1 \wedge \cdots \wedge \mathbf{x}_n. \quad [10.27]$$

By definition, we set that $I(u_1 \cdots u_n)$ is an invariant of the family $u_1, \cdots, u_n$.

PROOF.– The endomorphism $[u_1 \cdots u_n]$ belongs to $\mathcal{L}(\Lambda^n \boldsymbol{E})$, where $\Lambda^n \boldsymbol{E}$ has a dimension 1, hence the existence of the real in question.

Let us illustrate immediately an application:

PROPOSITION 10.4.– Let $u_1, \cdots, u_n$ and $v_1, \cdots, v_n$ be two families of n endomorphisms over $\boldsymbol{E}$. Then, we have:

$$I([u_1 \cdots u_n] \circ [v_1 \cdots v_n]) = I([u_1 \cdots u_n]) \, I([v_1 \cdots v_n]). \quad [10.28]$$

REMARK 10.1.– Before starting the demonstration, note that there is no simple rule for directly calculating the symbol $[u_1 \cdots u_n] \circ [v_1 \cdots v_n]$ as a function of u_i, v_j. The only particular situation that can be easily treated is when $u_i = u$ and $v_j = v$, in which case we obtain the formulas of proposition 10.2, point 1. ■

PROOF.– For any family $\mathbf{x}_1, \cdots, \mathbf{x}_n$, we have:

$$\begin{aligned} &[u_1 \cdots u_n] \circ [v_1 \cdots v_n](\mathbf{x}_1 \wedge \cdots \wedge \mathbf{x}_n) \\ &\quad = [u_1 \cdots u_n]\left(\frac{1}{n!} \sum_{\alpha \in S_n} v_{\alpha(1)}(\mathbf{x}_1) \wedge \cdots \wedge v_{\alpha(n)}(\mathbf{x}_n)\right). \end{aligned} \quad [10.29]$$

By linearity, we then have:

$$[u_1 \cdots u_n] \circ [v_1 \cdots v_n] (\mathbf{x}_1 \wedge \cdots \wedge \mathbf{x}_n)$$

$$= I(u_1 \cdots u_n) \left(\frac{1}{n!} \sum_{\alpha \in S_n} v_{\alpha(1)}(\mathbf{x}_1) \wedge \cdots \wedge v_{\alpha(n)}(\mathbf{x}_n) \right). \qquad [10.30]$$

We also find that:

$$\left(\frac{1}{n!} \sum_{\alpha \in S_n} v_{\alpha(1)}(\mathbf{x}_1) \wedge \cdots \wedge v_{\alpha(n)}(\mathbf{x}_n) \right) = [v_1 \cdots v_n] (\mathbf{x}_1 \wedge \cdots \wedge \mathbf{x}_n). \qquad [10.31]$$

Hence, finally,

$$[u_1 \cdots u_n] \circ [v_1 \cdots v_n] (\mathbf{x}_1 \wedge \cdots \wedge \mathbf{x}_n)$$

$$= I(u_1 \cdots u_n) I(v_1 \cdots v_n) \mathbf{x}_1 \wedge \cdots \wedge \mathbf{x}_n, \qquad [10.32]$$

and therefore, in virtue of the uniqueness of symbol I, we have the stated equality.

10.1.3. *Applications of invariants of an endomorphism*

We are going to employ the previous definitions and properties to study the invariants of an endomorphism:

DEFINITION 10.5.– Let u be an endomorphism over $\boldsymbol{E}$ and let k be an integer of $[0, n]$. We call invariant of degree k associated with u the real:

$$I_k(u) := \frac{n!}{k!(n-k)!} I(u \cdots u I_d \cdots I_d), \qquad [10.33]$$

where the endomorphism symbol u appears k times within the argument, whereas the identity operator appears $n - k$ times.

Let us remind that in practice:

DEFINITION 10.6.– The following specific terms are employed:

1) The invariant of degree 1 is called trace of the endomorphism, denoted by $Tr(u)$.

2) The invariant of degree n is called determinant of the endomorphism, denoted by $\det(u)$.

Furthermore, note that the invariant of degree 0 is always equal to 1.

PROPOSITION 10.5.– The following properties are true:

1) The determinant of an endomorphism is the unique real such that, for any family, we have:

$$u(\mathbf{x}_1) \wedge \cdots \wedge u(\mathbf{x}_n) = \det(u)\, \mathbf{x}_1 \wedge \cdots \wedge \mathbf{x}_n. \quad [10.34]$$

2) The determinant map $\det(\cdot)$ is homogeneous of degree n over the endomorphisms.

3) An endomorphism u is invertible if and only if $\det(u) \neq 0$.

4) The trace of an endomorphism is a linear operator $Tr(u+v) = Tr(u) + Tr(v)$.

5) For all endomorphism u, v, we have the equality: $I_k(u \circ v) = I_k(v \circ u)$.

PROOF.– Let us demonstrate these properties:

1) It suffices to consider the definitions.

2) Actually, we know that $[u]_n$ is invertible if and only if u is invertible (see proposition 10.2, point 2). Now, we know that $\det(u)$ is such that:

$$\forall \mathbf{E} \in \Lambda^n \boldsymbol{E}, \ [u]_n(\mathbf{E}) = \det(u)\, \mathbf{E}, \quad [10.35]$$

as a result, $[u]_n$ is invertible over $\Lambda^n \boldsymbol{E}$ if and only if $\det(u)$ is not null. Hence the result.

3) Immediate, since:

$$\begin{aligned} \det(\lambda u)[\mathbf{e}_1 \wedge \cdots \wedge \mathbf{e}_n] &= [\lambda u]_n [\mathbf{e}_1 \wedge \cdots \wedge \mathbf{e}_n] \\ &= (\lambda u)(\mathbf{e}_1) \wedge \cdots \wedge (\lambda u)(\mathbf{e}_n), \end{aligned} \quad [10.36]$$

and the latter term is equal to:

$$\lambda^n u(\mathbf{e}_1) \wedge \cdots \wedge u(\mathbf{e}_n) = \lambda^n \det(u)\, \mathbf{e}_1 \wedge \cdots \wedge \mathbf{e}_n. \quad [10.37]$$

4) It suffices to write that:

$$Tr(u+v)\, \mathbf{e}_1 \wedge \cdots \mathbf{e}_n = \sum_{i=1}^{i=n} \mathbf{e}_1 \wedge \cdots \wedge (u+v)(\mathbf{e}_i) \wedge \cdots \wedge \mathbf{e}_n, \quad [10.38]$$

the right-hand side can be developed instantly to obtain:

$$\sum_{i=1}^{i=n} \mathbf{e}_1 \wedge \cdots \wedge v(\mathbf{e}_i) \wedge \cdots \wedge \mathbf{e}_n + \sum_{i=1}^{i=n} \mathbf{e}_1 \wedge \cdots \wedge u(\mathbf{e}_i) \wedge \cdots \wedge \mathbf{e}_n, \quad [10.39]$$

which is thus equal to:

$$(Tr(v) + Tr(u))\,\mathbf{e}_1 \wedge \cdots \wedge \mathbf{e}_n, \qquad [10.40]$$

hence, we deduce the linearity of the trace operator.

5) Let us begin with the trace case: let us choose a basis $\mathbf{e}_1, \cdots, \mathbf{e}_n$ of $\boldsymbol{E}$. For two endomorphisms u, v, we have:

$$\sum_{i=1}^{i=n} \mathbf{e}_1 \wedge \cdots \wedge (v \circ u)(\mathbf{e}_i) \wedge \cdots \wedge \cdots \wedge \mathbf{e}_n. \qquad [10.41]$$

Within the expansion of $(v \circ u)(\mathbf{e}_i)$, we can keep only the term relative to $\mathbf{e}_i$. By introducing the matrix v_{ij}, u_{ij} representing the endomorphisms u, v with respect to the basis of $\mathbf{e}_i$, we immediately obtain that the component of $(v \circ u)$ over $\mathbf{e}_i$ can be written as $\sum_{k=1}^{k=n} v_{ik}u_{ki}$. Thus, we deduce that:

$$\sum_{i=1}^{i=n} \mathbf{e}_1 \wedge \cdots \wedge (v \circ u)(\mathbf{e}_i) \wedge \cdots \wedge \mathbf{e}_n = \left(\sum_{i=1}^{i=n}\sum_{k=1}^{k=n} v_{ik}u_{ki}\right)\mathbf{e}_1 \wedge \cdots \wedge \mathbf{e}_n, \qquad [10.42]$$

now, we can see instantly that the term that factorizes the n-blade is now symmetric with respect to u and v, which concludes our proof regarding the fact that $Tr(u \circ v) = Tr(v \circ u)$. As for the other invariants, the demonstration is identical:

$$\sum_{1 \le i_1 < \cdots < i_k \le n} \mathbf{e}_1 \wedge \cdots u \circ v(\mathbf{e}_{i_1}) \wedge \cdots \wedge u \circ v(\mathbf{e}_{i_k}) \wedge \cdots \wedge \mathbf{e}_n. \qquad [10.43]$$

The terms over the $\mathbf{e}_{i_p}, p \in [1, k]$ have to be developed only over the $\mathbf{e}_{i_r}, r \in [1, k]$. If $f : \boldsymbol{E} \mapsto \boldsymbol{E}$ is an endomorphism, we define $\widetilde{f}$ so that:

$$\forall r \in \{i_1, \cdots, i_k\}, \quad \widetilde{f}(\mathbf{e}_r) = \sum_{s \in \{i_1, \cdots, i_k\}} f_{rs}\mathbf{e}_s. \qquad [10.44]$$

Let σ be a permutation that maps $\{i_1, \cdots, i_k\}$ into $\{i_1, \cdots, i_k\}$. We look for the component of $\widetilde{u \circ v}(\mathbf{e}_{i_1})$ over the vector $\mathbf{e}_{\sigma(i_1)}$. Clearly, it equals:

$$\sum_{j_1 \in \{i_1, \cdots, i_k\}} v_{\sigma(i_1)j_1} u_{j_1 i_1} \mathbf{e}_{\sigma(i_1)}. \qquad [10.45]$$

Therefore, we obtain:

$$\sum_{1 \le i_1 < \cdots < i_k \le n} \sum_{\sigma \in SG_k} \left(\sum_{j_1 \in \{i_1, \cdots, i_k\}} v_{\sigma(i_1)j_1} u_{j_1 i_1}\right) \cdots \left(\sum_{j_k \in \{i_1, \cdots, i_k\}} v_{\sigma(i_k)j_k} u_{j_k i_k}\right) \varepsilon(\sigma)\,\mathbf{e}_1 \wedge \cdots \wedge \mathbf{e}_n \qquad [10.46]$$

with SG_k the subgroup of permutations of $\{1 \cdots n\}$ that maps $\{i_1, \cdots, i_k\}$ into itself while maintaining the other indexes fixed. However, it is clear that the term

$$\sum_{1 \le i_1 < \cdots < i_k \le n} \sum_{\sigma \in SG_k} \varepsilon(\sigma) \left(\sum_{j_1 \in \{i_1, \cdots, i_k\}} v_{\sigma(i_1) j_1} u_{j_1 i_1} \right) \cdots \left(\sum_{j_k \in \{i_1, \cdots, i_k\}} v_{\sigma(i_k) j_k} u_{j_k i_k} \right) \quad [10.47]$$

is symmetric over u and v (it is sufficient to perform a "change of variables" $\sigma \leftrightarrow \sigma^{-1}$ and then $\sigma^{-1} \leftrightarrow \sigma$ in order to "carry" the permutation terms from v to u). As a result, we have indeed the expected result.

Finally, an important theorem in linear algebra is illustrated:

THEOREM 10.3.– The characteristic polynomial of an endomorphism u is given by:

$$\forall X \in \mathbb{R}, \; P_u(X) := \det(u - XI_d) = \sum_{k=0}^{k=n} (-1)^k I_{n-k}(u) X^k. \quad [10.48]$$

PROOF.– It suffices to write that for all family $\mathbf{x}_1, \cdots, \mathbf{x}_n$, we have:

$$\det(u - XI_d)\, \mathbf{x}_1 \wedge \cdots \wedge \mathbf{x}_n = [u - XI_d]_n (\mathbf{x}_1 \wedge \cdots \wedge \mathbf{x}_n) = (u(\mathbf{x}_1) - X\mathbf{x}_1) \wedge \cdots \wedge (u(\mathbf{x}_n) - X\mathbf{x}_n). \quad [10.49]$$

Then, this expression is expanded and ordered according to the powers of $(-X)^k$. In order to construct an n-blade that is part of the coefficient of $(-X)^k$, it is necessary to choose the position $1 \le i_1 < \cdots < i_k \le n$ where we have a vector $\mathbf{e}_{i_k}$, with the other ones corresponding to a vector $u(\mathbf{e}_i)$. Therefore, it is possible to write that the factor of $(-X)^k$ can be written in the form:

$$\sum_{1 \le i_1 < \cdots < i_k \le n} u(\mathbf{e}_1) \wedge \cdots \mathbf{e}_{i_1} \wedge \cdots \wedge u(\mathbf{e}_{i_k}) \wedge \cdots \wedge \mathbf{e}_n. \quad [10.50]$$

It is straightforward that this term can be written exactly as:

$$k! [u \cdots u Id \cdots Id] (\mathbf{e}_1 \wedge \cdots \wedge \mathbf{e}_n) = k! I(u \cdots u I_d \cdots I_d)\, \mathbf{e}_1 \wedge \cdots \wedge \mathbf{e}_n, \quad [10.51]$$

where the symbol u appears $n-k$ times and the symbol I_d appears k times. Therefore, the result is demonstrated.

REMARK 10.2.– Note some interesting points:

1) The definition of endomorphism invariants involves the usage of symmetric brackets. This explains the fact that such invariants are always obtained *as summations*, and not directly as coefficients. In fact, for a fixed u and for a particular basis $\mathbf{e}_1, \cdots, \mathbf{e}_n$, we set:

$$\forall i \in [1, n], \quad u(\mathbf{e}_j) = \sum_{i=1}^{i=n} u_{ij} \mathbf{e}_i. \qquad [10.52]$$

Then, we can legitimately write that:

$$\mathbf{e}_1 \wedge \cdots \wedge u(\mathbf{e}_i) \wedge \cdots \wedge \mathbf{e}_n = u_{ii} \mathbf{e}_1 \wedge \cdots \wedge \mathbf{e}_n. \qquad [10.53]$$

If the left-hand side defined an endomorphism of $\Lambda^n \boldsymbol{E}$, then the coefficient of the right-hand side would be intrinsic: it would not depend on the chosen basis. This is not the case here, of course: the left-hand side is indeed *multi-linear* over $\boldsymbol{E}^n$, but it is *not skew-symmetric*. In order to make it skew-symmetric, it is necessary to sum over all the indexes i, which is precisely the role of theorem 10.1. Thus, we see that the sum of diagonal values of the matrix that represents an endomorphism is independent of the chosen basis: this is the trace of this endomorphism.

2) The considerations regarding invariants generally derive from considerations regarding similar matrices: recall that two matrices A, B (that are square, with a size n) are similar when there exists an invertible matrix P with a size n such that we have $B = P^{-1}AP$. The similarity relation is an equivalence relation. For a given matrix, it is always possible to consider A as a matrix that represents an endomorphism u with respect to a given basis. The equivalence class of A is then the set of matrices representing u in the set of constructible bases. So, considering the linear space of square matrices with a size n quotiented by the similarity relation, this space is exactly isomorphic to a set of endomorphisms. Posing the question of finding invariants with respect to similar matrices corresponds thus to posing the question of finding invariant quantities of endomorphisms. But then, we can extensively employ the space $\Lambda^n \boldsymbol{E}$, since all endomorphism of $\Lambda^n \boldsymbol{E}$ constructed from u necessarily returns an invariant quantity defined over u and incidentally over the set of matrices that represent the endomorphism. There is a great number of ways for constructing endomorphisms over $\Lambda^n \boldsymbol{E}$ starting from u. The problem consists then in knowing whether it is possible to construct methodically a complete system of invariants for an endomorphism (i.e. a set of invariants that characterizes this endomorphism, in other words: two endomorphisms are equal if and only if they have the same invariants) starting exclusively from the methods for constructing endomorphisms over $\Lambda^n \boldsymbol{E}$. Then, a naturally arising question is: given a complete system of invariants, how can the matrix of the endomorphism be calculated in whichever basis? ■

10.1.4. *Conjugated endomorphism*

Thanks to the duality operation, it is possible to transport endomorphisms of $\mathcal{L}(\Lambda^p \boldsymbol{E})$ into $\mathcal{L}(\Lambda^{n-p}\boldsymbol{E})$:

DEFINITION 10.7.– Let $F \in \mathcal{L}(\Lambda^p \boldsymbol{E})$. Then, we define its conjugated endomorphism $F^\natural \in \mathcal{L}(\Lambda^{n-p}\boldsymbol{E})$ by the relation:

$$\forall \mathbf{X} \in \Lambda^{n-p}\boldsymbol{E},\ \ F^\natural(\mathbf{X}) := \overline{F(\overline{\mathbf{X}})}. \qquad [10.54]$$

The following properties are corollaries that are immediately deduced from the definition:

COROLLARY 10.1.– The conjugation operator satisfies the following properties:

1) It is an involution:

$$\forall F \in \mathcal{L}(\Lambda^p \boldsymbol{E}),\ \ (F^\natural)^\natural = F. \qquad [10.55]$$

2) It is a linear operator:

$$\forall F, G \in \mathcal{L}(\Lambda^p \boldsymbol{E}), \forall \alpha \in \mathbb{R},\ \ (F + \alpha G)^\natural = F^\natural + \alpha G^\natural. \qquad [10.56]$$

3) The following formula holds:

$$\forall \mathbf{X}, \mathbf{Y} \in \Lambda^{n-p}\boldsymbol{E},\ \ \forall F, G \in \mathcal{L}(\Lambda^p \boldsymbol{E}),$$
$$B_{n-p}\left(F^\natural(\mathbf{X}), G^\natural(\mathbf{Y})\right) = (-1)^s B_p\left(F(\overline{\mathbf{X}}), G(\overline{\mathbf{Y}})\right). \qquad [10.57]$$

4) The conjugation operator commutes with the (adjoint) transposition operator with respect to the dot product.

5) F is invertible if and only if $F^\natural$ is invertible. The conjugation commutes with the inversion. We have:

$$\forall F \in \mathcal{L}(\Lambda^p \boldsymbol{E}),\ \ \det\left(F^\natural\right) = (-1)^{N(p(n-p)+s)} \det(F), \qquad [10.58]$$

where $N = \binom{n}{p} = n!/(p!(n-p)!)$ is the dimension of the space $\Lambda^p \boldsymbol{E}$.

6) F is an isometry over $\Lambda^p \boldsymbol{E}$ if and only if $F^\natural$ is an isometry over $\Lambda^{n-p}\boldsymbol{E}$.

PROOF.– Let us demonstrate these properties:

1) We have:

$$\forall \mathbf{X} \in \Lambda^p \boldsymbol{E}, \quad F^{\natural\natural}(\mathbf{X}) =$$
$$\overline{\overline{F\left(\overline{\overline{\mathbf{X}}}\right)}} = (-1)^{p(n-p)+s}(-1)^{p(n-p)+s} F(\mathbf{X}) = F(\mathbf{X}). \quad [10.59]$$

2) Evident since the duality operator is itself linear.

3) We employ the fact that the conjugation is a signed pseudo-isometry (see theorem 9.2):

$$\forall \mathbf{E}, \mathbf{F}, \in \Lambda^p \boldsymbol{E}, \quad B_{n-p}\left(\overline{\mathbf{E}}, \overline{\mathbf{F}}\right) = (-1)^s B_p(\mathbf{E}, \mathbf{F}). \quad [10.60]$$

We apply this both for $\mathbf{E} = \overline{F\left(\overline{\mathbf{X}}\right)}$ and $\mathbf{F} = \overline{G\left(\overline{\mathbf{Y}}\right)}$. We deduce thereof:

$$B_{n-p}\left(F^\natural(\mathbf{X}), G^\natural(\mathbf{Y})\right) = (-1)^s B_p\left(F\left(\overline{\mathbf{X}}\right), G\left(\overline{\mathbf{Y}}\right)\right). \quad [10.61]$$

4) For all $\mathbf{X}, \mathbf{Y}$ of $\Lambda^{n-p}\boldsymbol{E}$, it is possible to write:

$$B_{n-p}\left(F^\natural(\mathbf{X}), \mathbf{Y}\right) = B_{n-p}\left(\overline{F\left(\overline{\mathbf{X}}\right)}, \mathbf{Y}\right)$$
$$= (-1)^{p(n-p)+s} B_{n-p}\left(\overline{F\left(\overline{\mathbf{X}}\right)}, \overline{\overline{\mathbf{Y}}}\right). \quad [10.62]$$

Then, we apply the previously demonstrated equality and we obtain:

$$= (-1)^{p(n-p)} B_p\left(F\left(\overline{\mathbf{X}}\right), \overline{\mathbf{Y}}\right) = (-1)^{p(n-p)} B_p\left(\overline{\mathbf{X}}, F^*\left(\overline{\mathbf{Y}}\right)\right). \quad [10.63]$$

We use now theorem 9.2 to write:

$$= (-1)^{p(n-p)}(-1)^s B_{n-p}\left(\overline{\overline{\mathbf{X}}}, \overline{F^*\left(\overline{\mathbf{Y}}\right)}\right). \quad [10.64]$$

Then, we employ proposition 9.3 to conclude:

$$= B_{n-p}\left(\mathbf{X}, \overline{F^*\left(\overline{\mathbf{Y}}\right)}\right) = B_{n-p}\left(\mathbf{X}, [F^*]^\natural(\mathbf{Y})\right), \quad [10.65]$$

hence, the fact that $[F^\natural]^* = [F^*]^\natural$.

5) Actually, we can solve:

$$F^\natural(\mathbf{X}) = \mathbf{Y} \Leftrightarrow \overline{F\left(\overline{\mathbf{X}}\right)} = \mathbf{Y} \Leftrightarrow F\left(\overline{\mathbf{X}}\right) = (-1)^{p(n-p)+s}\,\overline{\mathbf{Y}} \Leftrightarrow$$
$$\overline{\mathbf{X}} = (-1)^{p(n-p)+s} F^{-1}\left(\overline{\mathbf{Y}}\right), \quad [10.66]$$

in other words,

$$F^{\natural}(\mathbf{X}) = \mathbf{Y} \Leftrightarrow \mathbf{X} = \overline{F^{-1}\left(\overline{\mathbf{Y}}\right)}, \qquad [10.67]$$

Hence, the fact that we have indeed $\left[F^{\natural}\right]^{-1} = \left[F^{-1}\right]^{\natural}$. To prove the formula of the determinant calculation, consider a basis $\mathbf{E}_i, i \in [1, N]$ of the space $\Lambda^p \boldsymbol{E}$. We write:

$$F(\mathbf{E}_i) = \sum_{j=1}^{j=N} F_{ij}\mathbf{E}_j \Rightarrow \overline{F\left(\overline{\overline{\mathbf{E}_i}}\right)} = (-1)^{p(n-p)+s} \sum_{j=1}^{j=n} F_{ij}\overline{\mathbf{E}_j}. \qquad [10.68]$$

On the other hand, we have:

$$\overline{F\left(\overline{\overline{\mathbf{E}_i}}\right)} := F^{\natural}\left(\overline{\mathbf{E}_i}\right) := \sum_{j=1}^{j=N} F^{\natural}_{ij}\overline{\mathbf{E}_j}. \qquad [10.69]$$

Hence, we deduce:

$$\sum_{j=1}^{j=N} F^{\natural}_{ij}\overline{\mathbf{E}_j} = (-1)^{p(n-p)+s} \sum_{j=1}^{j=n} F_{ij}\overline{\mathbf{E}_j}. \qquad [10.70]$$

By identification, we have $F^{\natural}_{ij} = (-1)^{p(n-p)+s} F_{ij}$. Thus, by homogeneity of the determinant function, we have:

$$\det\left(F^{\natural}\right) = (-1)^{N(p(n-p)+s)} \det(F), \qquad [10.71]$$

where N is the dimension of the space $\Lambda^p \boldsymbol{E}$, in other words, $N = \binom{n}{p} = n!/(p!(n-p)!)$.

6) This is a consequence of the two previous properties, since an isometry u is characterized by $u^{-1} = u^*$.

10.2. Decomposition of endomorphisms of exterior algebras

An interesting algebraic problem consists of trying to recognize constructible endomorphisms, or, more precisely, in trying to define the reciprocal process of endomorphism construction of $\Lambda^p \boldsymbol{E}$ via the symmetric bracket:

> QUESTION 10.2.– Given $F : \Lambda^p \boldsymbol{E} \mapsto \Lambda^p \boldsymbol{E}$, are there any endomorphisms $f_1, \cdots, f_p$ over $\boldsymbol{E}$ such that it is possible to construct F from f_i in the form $F = [f_1 \cdots f_p]$?

In its general sense, this question is far beyond the scope of this work. In practice, we shall restrict ourselves to the case $p = n - 1$ in which all the endomorphisms f_i are equal: therefore, we are trying to factorize, if possible, endomorphisms $F \in \mathcal{L}\left(\Lambda^{n-1}\boldsymbol{E}\right)$ in the form $F = [f]_{n-1}$.

10.2.1. *Examination of* $\mathcal{L}\left(\Lambda^2\mathbb{R}^3\right)$

In the case of dimension three, it is possible to pose the problem presented above in a particular way. In fact, let us consider a given F, described by its components with respect to a basis, and let us find f with which F can be constructed:

$$\begin{cases} F(\mathbf{e}_1 \wedge \mathbf{e}_2) = F_{12,12}\mathbf{e}_1 \wedge \mathbf{e}_2 + F_{12,13}\mathbf{e}_1 \wedge \mathbf{e}_3 + F_{12,23}\mathbf{e}_2 \wedge \mathbf{e}_3 \\ F(\mathbf{e}_1 \wedge \mathbf{e}_3) = F_{13,12}\mathbf{e}_1 \wedge \mathbf{e}_2 + F_{13,13}\mathbf{e}_1 \wedge \mathbf{e}_3 + F_{13,23}\mathbf{e}_2 \wedge \mathbf{e}_3 \\ F(\mathbf{e}_2 \wedge \mathbf{e}_3) = F_{23,12}\mathbf{e}_1 \wedge \mathbf{e}_2 + F_{23,13}\mathbf{e}_1 \wedge \mathbf{e}_3 + F_{23,23}\mathbf{e}_2 \wedge \mathbf{e}_3 \end{cases} \quad [10.72]$$

$$\begin{cases} f(\mathbf{e}_1) = f_{1,1}\mathbf{e}_1 + f_{1,2}\mathbf{e}_2 + f_{1,3}\mathbf{e}_3 \\ f(\mathbf{e}_2) = f_{2,1}\mathbf{e}_1 + f_{2,2}\mathbf{e}_2 + f_{2,3}\mathbf{e}_3 \\ f(\mathbf{e}_3) = f_{3,1}\mathbf{e}_1 + f_{3,2}\mathbf{e}_2 + f_{3,3}\mathbf{e}_3 \end{cases} \quad [10.73]$$

If f allows us to construct F, it is necessary to identify $f(\mathbf{e}_i) \wedge f(\mathbf{e}_j)$ with respect to $F(\mathbf{e}_i \wedge \mathbf{e}_i)$, which actually yields *a priori* nine equations with nine unknowns:

$$\begin{bmatrix} (f_{11}f_{22} - f_{21}f_{12}) & (f_{11}f_{23} - f_{21}f_{13}) & (f_{12}f_{23} - f_{22}f_{13}) \\ (f_{11}f_{32} - f_{31}f_{12}) & (f_{11}f_{33} - f_{31}f_{13}) & (f_{12}f_{33} - f_{32}f_{13}) \\ (f_{21}f_{32} - f_{31}f_{22}) & (f_{21}f_{33} - f_{31}f_{33}) & (f_{22}f_{33} - f_{32}f_{23}) \end{bmatrix} \quad [10.74]$$

which has to be equal, term by term, to:

$$\begin{bmatrix} F_{12,12} & F_{12,13} & F_{12,23} \\ F_{13,12} & F_{13,13} & F_{13,23} \\ F_{23,12} & F_{23,13} & F_{23,23} \end{bmatrix}. \quad [10.75]$$

An evident difficulty of this system is that it is quite nonlinear. As a result, without a specific interpretation, it is unclear how we could answer the question of a solution's existence, and even less clear how this solution can be found explicitly. Introducing the conjugation will allow us to interpret the solution problem of this linear system as a problem of matrix inversion. In fact, let us set $g := F^{\natural}$. Since the conjugation is an involution, we can easily write that we have $F = g^{\natural}$. Then, we can rewrite the components of F within the basis $\mathbf{e}_i \wedge \mathbf{e}_j$ starting from the components of g in the basis of e_i, obtaining thus:

$$\begin{aligned} F(\mathbf{e}_1 \wedge \mathbf{e}_2) &= g_{3,3}\mathbf{e}_1 \wedge \mathbf{e}_2 - g_{3,2}\mathbf{e}_1 \wedge \mathbf{e}_3 + g_{3,1}\mathbf{e}_2 \wedge \mathbf{e}_3 \\ F(\mathbf{e}_1 \wedge \mathbf{e}_3) &= -g_{2,3}\mathbf{e}_1 \wedge \mathbf{e}_2 + g_{2,2}\mathbf{e}_1 \wedge \mathbf{e}_3 - g_{2,1}\mathbf{e}_2 \wedge \mathbf{e}_3 \\ F(\mathbf{e}_2 \wedge \mathbf{e}_3) &= g_{1,3}\mathbf{e}_1 \wedge \mathbf{e}_2 - g_{1,2}\mathbf{e}_1 \wedge \mathbf{e}_3 + g_{1,1}\mathbf{e}_2 \wedge \mathbf{e}_3 \end{aligned} \quad [10.76]$$

The above matrix equality can be rewritten in the form:

$$\begin{bmatrix} (f_{11}f_{22}-f_{21}f_{12}) & (f_{11}f_{23}-f_{21}f_{13}) & (f_{12}f_{23}-f_{22}f_{13}) \\ (f_{11}f_{32}-f_{31}f_{12}) & (f_{11}f_{33}-f_{31}f_{13}) & (f_{12}f_{33}-f_{32}f_{13}) \\ (f_{21}f_{32}-f_{31}f_{22}) & (f_{21}f_{33}-f_{31}f_{33}) & (f_{22}f_{33}-f_{32}f_{23}) \end{bmatrix} \qquad [10.77]$$

which has to be equal, term by term, to:

$$\begin{bmatrix} g_{3,3} & -g_{3,2} & g_{3,1} \\ -g_{2,3} & g_{2,2} & -g_{2,1} \\ g_{1,3} & -g_{1,2} & g_{1,1} \end{bmatrix}. \qquad [10.78]$$

The equalities are re-arranged term by term by bringing up the matrix of the operator g in the basis of the e_i at the right-hand side. Thus, we obtain the equality:

$$\begin{bmatrix} (f_{22}f_{33}-f_{32}f_{23}) & -(f_{21}f_{33}-f_{31}f_{33}) & (f_{21}f_{32}-f_{31}f_{22}) \\ -(f_{12}f_{33}-f_{32}f_{13}) & (f_{11}f_{33}-f_{31}f_{13}) & -(f_{11}f_{32}-f_{31}f_{12}) \\ (f_{12}f_{23}-f_{22}f_{13}) & -(f_{11}f_{23}-f_{21}f_{13}) & (f_{11}f_{22}-f_{21}f_{12}) \end{bmatrix} \qquad [10.79]$$

that equals term by term:

$$\begin{bmatrix} g_{1,1} & g_{1,2} & g_{1,3} \\ g_{2,1} & g_{2,2} & g_{2,3} \\ g_{3,1} & g_{3,2} & g_{3,3} \end{bmatrix}. \qquad [10.80]$$

At this point, it seems that the re-arrangement is miraculous: in the left-hand side, we recognize the matrix of co-factors that are associated with the representative matrix of f in the basis of $\mathbf{e}_i$, meanwhile the right-hand side matrix is the matrix of g. Therefore, we have to solve the equation:

$$Com\,(f) = M\,(g), \qquad [10.81]$$

where $M\,(g)$ is the matrix of g in the basis of $\mathbf{e}_i$, which, let us remind it, is orthonormal. By transposing the equality and by introducing the adjoint g^* of g, we obtain exactly:

$$^{t}Com\,(f) = M\,(g^*). \qquad [10.82]$$

If g is invertible, then g^* is also invertible and $M\,(g^*)$ is invertible as a matrix. According to the previous equality, $^{t}Com\,(f)$ is also an invertible matrix. We remind Laplace's formula: if f is invertible, its representative matrix in whichever basis is also invertible. Furthermore, the inverse matrix, the matrix of the operator's inverse and the matrix of co-factors are related by the equalities:

$$M\left(f^{-1}\right) = M^{-1}(f) = \frac{1}{\det(f)}\,{}^{t}Com(f). \qquad [10.83]$$

Thus, if the matrix of co-factors of a map f is invertible, this means that the operator f is itself invertible. Incidentally, note that by considering the determinant of the two last terms, we reach the conclusion that:

$$\det\left({}^{t}Com(f)\right) = [\det(f)]^{2}. \qquad [10.84]$$

As a result, equality [10.82] implies that, in order to find a solution f to the construction problem, we need to have necessarily:

$$\det(g) = \det F^{\natural} > 0. \qquad [10.85]$$

We suppose that this is indeed the case. Therefore, by considering Laplace's equalities, if g is invertible, then f is invertible. By considering the inverse of equality [10.82], we obtain exactly:

$${}^{t}Com\left(f^{-1}\right) = M\left(g^{*-1}\right). \qquad [10.86]$$

In other words,

$$\frac{1}{\det(f)}M(f) = M\left(g^{*-1}\right) \Leftrightarrow f = \det(f)\,[g^{*}]^{-1} = \frac{1}{\det(g)^{1/2}}\,[g^{*}]^{-1}. \qquad [10.87]$$

However, reciprocally, let us suppose that $\det(g) > 0$ and consider now:

$$f := \frac{1}{\det(g)^{1/2}}\,[g^{*}]^{-1}, \qquad [10.88]$$

the question then consists of knowing whether f enables the construction of F, namely, do we have $F = [f]_2$? The answer is positive, as we shall demonstrate in a more general theorem.

10.2.2. *Constructibility of endomorphisms of* $\Lambda^{n-1}\boldsymbol{E}$

THEOREM 10.4.– Let $\boldsymbol{E}$ be a linear space with a dimension n and let $F : \Lambda^{n-1}\boldsymbol{E} \mapsto \Lambda^{n-1}\boldsymbol{E}$ be an isomorphism over $\Lambda^{n-1}\boldsymbol{E}$ such that $\det(F)$ belongs to the image of the function $\left[x \in \mathbb{R},\ x \mapsto (-1)^s x^{n-1}\right]$. Then, there exists a unique map $f \in \mathcal{L}(E)$ that allows us to construct F. This map is invertible and can be written in the form:

$$f = \left[(-1)^{n-1+s}\det(F)\right]^{1/n-1}\left((-1)^{n-1+s}F^{\natural *}\right)^{-1}, \qquad [10.89]$$

where the symbol $\natural$ designates the conjugation operator, meanwhile the symbol $*$ designates the transposition operator (i.e. the adjoint) for the pseudo-dot product.

PROOF.– Suppose that F is invertible and let us find f such that $F = [f]_{n-1}$. For every pair $\mathbf{x}, \mathbf{y} \in \boldsymbol{E}$, we have:

$$F(\overline{\mathbf{y}}) \wedge f(\mathbf{x}) = [f]_{n-1}(\overline{\mathbf{y}}) \wedge f(\mathbf{x}). \qquad [10.90]$$

If $\mathbf{y}$ is a vector, its conjugate $\overline{\mathbf{y}}$ is an element of $\Lambda^{n-1}\boldsymbol{E}$ and therefore, it is decomposable, i.e. it can be written as an $n-1$-blade (see proposition 8.7). Thus, it is written in the form $\mathbf{y}_1 \wedge \cdots \wedge \mathbf{y}_{n-1}$, and we can write:

$$[f]_{n-1}(\overline{\mathbf{y}}) = f(\mathbf{y}_1) \wedge \cdots \wedge f(\mathbf{y}_{n-1}). \qquad [10.91]$$

This allows us to follow up with:

$$[f]_{n-1}(\overline{\mathbf{y}}) \wedge f(\mathbf{x}) = f(\mathbf{y}_1) \wedge \cdots \wedge f(\mathbf{y}_{n-1}) \wedge f(\mathbf{x}) = \det(f)\,\overline{\mathbf{y}} \wedge \mathbf{x}. \qquad [10.92]$$

Now, according to proposition 9.9, we know that:

$$\overline{\mathbf{y}} \wedge \mathbf{x} = b(\mathbf{y}, \mathbf{x})\,\mathbf{I}. \qquad [10.93]$$

Hence, we deduce:

$$F(\overline{\mathbf{y}}) \wedge f(\mathbf{x}) = \det(f)\,b(\mathbf{y}, \mathbf{x})\,\mathbf{I}. \qquad [10.94]$$

We can consider the H-conjugate of each term; at the right-hand side, we obtain:

$$\overline{\det(f)\,b(\mathbf{y}, \mathbf{x})\,\mathbf{I}} = \det(f)\,b(\mathbf{y}, \mathbf{x})\,\overline{\mathbf{I}} = (-1)^s \det(f)\,b(\mathbf{y}, \mathbf{x}), \qquad [10.95]$$

since $\overline{\mathbf{I}} = (-1)^s$. On the other hand, the conjugate of the left-hand side of formula [10.90] can be written as:

$$\overline{F(\overline{\mathbf{y}}) \wedge f(\mathbf{x})} = (-1)^s \overline{F(\overline{\mathbf{y}})} \vee \overline{f(\mathbf{x})} = (-1)^s F^\natural(\mathbf{y}) \vee \overline{f(\mathbf{x})} = (-1)^s (-1)^{n-1} \overline{f(\mathbf{x})} \vee F^\natural(\mathbf{y}). \quad [10.96]$$

The latter equality being obtained via proposition 9.7. Furthermore, by using proposition 9.9, we have:

$$\overline{f(\mathbf{x})} \vee F^\natural(\mathbf{y}) = (-1)^s b\left(F^\natural(\mathbf{y}), f(\mathbf{x})\right). \quad [10.97]$$

Thus, the following equality holds for every pair $\mathbf{x}, \mathbf{y}$:

$$(-1)^{n-1} b\left(F^\natural(\mathbf{y}), f(\mathbf{x})\right) = (-1)^s \det(f)\, b(\mathbf{y}, \mathbf{x}) \Leftrightarrow b\left(\mathbf{y}, \left[F^\natural\right]^* (f(\mathbf{x}))\right) = (-1)^{n-1+s} \det(f)\, b(\mathbf{x}, \mathbf{y}). \quad [10.98]$$

Since $b(\cdot,\cdot)$ is a nondegenerate form, it entails that:

$$\forall \mathbf{x} \in \boldsymbol{E}, F^{\natural *}(f(\mathbf{x})) = (-1)^{n-1+s} \det(f)\, \mathbf{x}. \quad [10.99]$$

As we supposed F to be invertible, if f constructs F, f necessarily has to be invertible (see proposition 10.2, point 2) and thus $\det(f) \neq 0$. By considering the determinant of both sides of the equation, we reach exactly:

$$\det\left(F^{\natural *}\right) \det(f) = (-1)^{n(n-1+s)} [\det(f)]^n \Rightarrow \det\left(F^\natural\right) = (-1)^{n(n-1+s)} [\det(f)]^{n-1}, \quad [10.100]$$

in other words, since $\det(F^\natural) = (-1)^{(n-1)(n-1+s)} \det(F)$ according to proposition 10.1, point 5, we know that we have:

$$(-1)^{n(n-1+s)} \det\left(F^\natural\right) = [\det(f)]^{n-1} \Leftrightarrow (-1)^{n-1+s} \det(F) = [\det(f)]^{n-1}. \quad [10.101]$$

We also know that $\det(F) = (-1)^s a^{n-1}$ (hypothesis of the theorem) for a certain real a. Therefore, it is possible to invert the previous formula:

$$\det(f) = \left[(-1)^{n-1+s} \det(F)\right]^{1/n-1}. \quad [10.102]$$

Equation [10.99] can thus be rewritten with the only unknown f and other (known) terms related to $F^{\natural}$, in the form:

$$\forall \mathbf{x} \in \boldsymbol{E},\ \ F^{\natural *}\left(f\left(\mathbf{x}\right)\right) = \left(-1\right)^{n-1+s}\left[\left(-1\right)^{n-1+s}\det\left(F\right)\right]^{1/n-1}\mathbf{x}. \qquad [10.103]$$

We see that f appears naturally as an inverse:

$$f = \left[\left(-1\right)^{n-1+s}\det\left(F\right)\right]^{1/n-1}\left(\left(-1\right)^{n-1+s}F^{\natural *}\right)^{-1}, \qquad [10.104]$$

which actually exists, since F and thus $F^{\natural}$ and also $F^{\natural *}$ are invertible. Consequently, when $F \in \mathcal{L}\left(\Lambda^{n-1}\boldsymbol{E}\right)$ is invertible with a determinant within the set $\left\{x^{n-1}, x \in \mathbb{R}\right\}$, if f enables the construction of F, f is necessarily invertible, and it exists in a unique way in virtue of formula [10.104]. Now, it is necessary to verify that such f actually allows us to construct F. Before starting with the calculations, note that we have demonstrated that the operators $\natural, *, -1$ commute when they are applied to an endomorphism $F \in \mathcal{L}\left(\Lambda^{p}\boldsymbol{E}\right)$ (in general terms, we already know that $*, -1$ commute and we have demonstrated that $*, \natural$ and $\natural, -1$ also commute). Therefore, to simplify the notations, it is possible to juxtapose these symbols and to exchange their order. So, we calculate the following term for all $\mathbf{x}, \mathbf{y} \in \boldsymbol{E}$:

$$\left[f\right]_{n-1}\left(\overline{\mathbf{y}}\right) \wedge F^{*-1\natural}\left(\mathbf{x}\right). \qquad [10.105]$$

By expanding the symmetric bracket $\left[\cdot\right]_{n-1}$, it can be seen that it is equal to:

$$\left[\left(-1\right)^{n-1+s}\left[\left(-1\right)^{n-1+s}\det\left(F\right)\right]^{1/n-1}\left(F^{\natural *}\right)^{-1}\right]_{n-1}\left(\overline{\mathbf{y}}\right) \wedge F^{*-1\natural}\left(\mathbf{x}\right), \qquad [10.106]$$

which in turn is equal to:

$$\left(-1\right)^{n(n-1+s)}\det\left(F\right)\left[F^{\natural *-1}\right]_{n-1}\left(\overline{\mathbf{y}}\right) \wedge F^{*-1\natural}\left(\mathbf{x}\right). \qquad [10.107]$$

Now, by reminding that we have $\det\left(F^{\natural}\right) = \left(-1\right)^{(n-1)(n-1+s)}\det\left(F\right)$, this last term equals:

$$= \left(-1\right)^{n-1+s}\det\left(F^{\natural}\right)\det\left(F^{\natural *-1}\right)\left(\overline{\mathbf{y}} \wedge \mathbf{x}\right) = \left(-1\right)^{n-1+s}\left(\overline{\mathbf{y}} \wedge \mathbf{x}\right), \qquad [10.108]$$

let thus:

$$\left[f\right]_{n-1}\left(\overline{\mathbf{y}}\right) \wedge F^{*-1\natural}\left(\mathbf{x}\right) = \left(-1\right)^{n-1+s}\left(\overline{\mathbf{y}} \wedge \mathbf{x}\right) = \left(-1\right)^{n-1+s}b\left(\mathbf{y}, \mathbf{x}\right)\mathbf{I}. \quad [10.109]$$

By considering the dual of the equality above, we obtain:

$$\overline{[f]_{n-1}(\overline{\mathbf{y}}) \wedge F^{*-1\natural}(\mathbf{x})} = (-1)^{n-1+s}(-1)^s b(\mathbf{y},\mathbf{x}). \qquad [10.110]$$

In other words, according to proposition 9.2:

$$\overline{[f]_{n-1}(\overline{\mathbf{y}})} \vee \overline{F^{*-1\natural}(\mathbf{x})} = (-1)^{n-1+s} b(\mathbf{y},\mathbf{x}), \qquad [10.111]$$

and therefore, thanks to proposition 9.9, we have indeed:

$$(-1)^{n-1+s} b\left([f]_{n-1}^{\natural}(\mathbf{y}), F^{*-1\natural}(\mathbf{x})\right) = (-1)^{n-1+s} b(\mathbf{y},\mathbf{x}). \qquad [10.112]$$

As this equality holds for all $\mathbf{x}, \mathbf{y}$, it allows us to write:

$$\forall \mathbf{y} \in \boldsymbol{E},\ F^{-1\natural}\left([f]_{n-1}^{\natural}(\mathbf{y})\right) = \mathbf{y} \Leftrightarrow$$
$$\forall \mathbf{y} \in \boldsymbol{E},\ [f]_{n-1}^{\natural}(\mathbf{y}) = F^{\natural}(\mathbf{y}) \Leftrightarrow [f]_{n-1} = F, \qquad [10.113]$$

which is exactly what we were seeking to demonstrate.

10.2.3. *Laplace inversion formula*

THEOREM 10.5 (Laplace inversion formula).– Let $f \in \mathcal{L}(\boldsymbol{E})$ be invertible, with $(\boldsymbol{E}, b)$ a pseudo-Euclidean space with a signature s. Then, it is possible to calculate its inverse operator via the Laplace formula:

$$f^{-1} = \frac{(-1)^{n-1+s}}{\det(f)} [f]_{n-1}^{\natural *}. \qquad [10.114]$$

PROOF.– Immediate: if $F = [f]_{n-1}$ with f invertible, then since $\det F = (-1)^{n-1+s}(\det f)^{n-1}$ (according to the calculations presented in the above section), we have the hypotheses of the previous theorem 10.4. The endomorphism f is then the unique constructor of F and according to the previous theorem, we have:

$$f = \left[(-1)^{n-1+s}\det(F)\right]^{1/n-1}\left((-1)^{n-1+s} F^{\natural *}\right)^{-1}. \qquad [10.115]$$

Then, it is possible to invert the relation in order to obtain:

$$f^{-1} = \frac{(-1)^{n-1+s}}{\left[(-1)^{n-1+s}\det(F)\right]^{1/n-1}} F^{\natural *}. \qquad [10.116]$$

Furthermore, we also know that (see previous calculations):

$$(-1)^{n-1+s} \det(F) = [\det(f)]^{(n-1)}, \qquad [10.117]$$

and therefore:

$$f^{-1} = \frac{(-1)^{n-1+s}}{\det(f)} F^{\natural *} = \frac{(-1)^{n-1+s}}{\det(f)} [f]^{\natural *}_{n-1}. \qquad [10.118]$$

11

$\Lambda^2 \boldsymbol{E}$ Algebra

An important issue in the context of pseudo-Euclidean spaces $(\boldsymbol{E}, b)$ consists of identifying (pseudo-) isometries. For example, in physics, this problem underlies the whole discourse on Galilean mechanics (a Euclidean case) and relativistic mechanics (Minkowski's case, which is pseudo-Euclidean). In any case, the (pseudo-) isometries form a Lie group (i.e. a group equipped with a variety structure) with an associated Lie algebra, which is an algebra of (pseudo-) skew-symmetric operators. Since we encounter the words "map" and "skew-symmetric", the following question naturally arises:

QUESTION 11.1.– Is there any relation between pseudo-skew-symmetric operators of $(\boldsymbol{E}, b)$ and exterior algebras? In the positive case, how can the calculations within exterior algebras provide information regarding the structure of pseudo-skew- symmetric operators, and vice versa?

11.1. Correspondence between a skew-symmetric operator and $\Lambda^2 \boldsymbol{E}$ elements

11.1.1. *Elementary reminders on operators' symmetries*

DEFINITION 11.1.– Let $(\boldsymbol{E}, b)$ be a pseudo-Euclidean space and let $G : \boldsymbol{E} \mapsto \boldsymbol{E}$ be a linear operator over $\boldsymbol{E}$. The operator G is called:

1) Skew-symmetric when:

$$\forall \mathbf{x}, \mathbf{y} \in \boldsymbol{E}, \;\; b(G(\mathbf{x}), \mathbf{y}) = -b(\mathbf{x}, G(\mathbf{y})). \qquad [11.1]$$

2) Symmetric when:

$$\forall \mathbf{x}, \mathbf{y} \in \boldsymbol{E}, \quad b(G(\mathbf{x}), \mathbf{y}) = (\mathbf{x}, G(\mathbf{y})). \qquad [11.2]$$

Finally, we denote by $G \in \mathcal{A}(\boldsymbol{E})$ if G is skew-symmetric and by $G \in \mathcal{S}(\boldsymbol{E})$ if G is symmetric.

PROPOSITION 11.1.– Let $(\boldsymbol{E}, b)$ be a pseudo-Euclidean space. A linear operator $G : \boldsymbol{E} \mapsto \boldsymbol{E}$ is skew-symmetric if and only if it satisfies:

$$\forall \mathbf{x} \in \boldsymbol{E}, \; b(G(\mathbf{x}), \mathbf{x}) = 0. \qquad [11.3]$$

PROOF.– If G is skew-symmetric, we clearly have:

$$b(G(\mathbf{x}), \mathbf{x}) = -b(\mathbf{x}, G(\mathbf{x})) \Rightarrow b(G(\mathbf{x}), \mathbf{x}) = 0. \qquad [11.4]$$

Vice versa, let us consider $\mathbf{x}, \mathbf{y}$, some elements of $\boldsymbol{E}$, we have:

$$b(G(\mathbf{x}+\mathbf{y}), (\mathbf{x}+\mathbf{y})) = 0 \Rightarrow$$
$$b(G(\mathbf{x}), \mathbf{x}) + b(G(\mathbf{y}), \mathbf{y}) + b(G(\mathbf{x}), \mathbf{y}) + b(G(\mathbf{y}), \mathbf{x}) = 0, \qquad [11.5]$$

and thus $b(G(\mathbf{x}), \mathbf{y}) + b(G(\mathbf{y}), \mathbf{x}) = 0$ (since by hypothesis we know that $b(G(\mathbf{x}), \mathbf{x}) = b(G(\mathbf{y}), \mathbf{y}) = 0$), which proves that the operator G is skew-symmetric.

PROPOSITION 11.2.– Let $G : \boldsymbol{E} \mapsto \boldsymbol{E}$ be a skew-symmetric operator. Then, $G^2 := G \circ G$ is a symmetric operator over $\boldsymbol{E}$.

PROOF.– Evidently, it is enough to observe that:

$$b\left(G^2(\mathbf{x}), \mathbf{y}\right) = -b(G(\mathbf{x}), G(\mathbf{y})) = b\left(\mathbf{x}, G^2(\mathbf{y})\right), \qquad [11.6]$$

and G^2 satisfies, of course, the definition of symmetric operators.

PROPOSITION 11.3.– Let $G \in \mathcal{A}(\boldsymbol{E}) \cup \mathcal{S}(\boldsymbol{E})$. Then, we have $Im(G) = [\ker(G)]^{\perp}$.

PROOF.– Let $\mathbf{x} \in \ker(G)$. It is clear that:

$$\forall \mathbf{y} \in \boldsymbol{E}, \quad b(\mathbf{x}, G(\mathbf{y})) = b(G^*(\mathbf{x}), \mathbf{y}) = \pm b(G(\mathbf{x}), \mathbf{y}) = 0, \qquad [11.7]$$

which proves that $Im(G) \subset [\ker(G)]^{\perp}$. However, according to Grassmann's formula, we also have:

$$\dim(\boldsymbol{E}) = \dim(Im(G)) + \dim(\ker(G)) \Leftrightarrow$$
$$\dim(Im(G)) = n - \dim(\ker(G)), \quad [11.8]$$

and we also know, in virtue of proposition 6.5, that:

$$\dim\left([\ker(G)]^{\perp}\right) = n - \dim(\ker(G)), \quad [11.9]$$

hence, we finally deduce that:

$$Im(G) \subset [\ker(G)]^{\perp} \;\; \textit{and} \;\; \dim(Im(G)) = \dim\left([\ker(G)]^{\perp}\right), \quad [11.10]$$

which indeed implies $Im(G) = [\ker(G)]^{\perp}$.

PROPOSITION 11.4.– Let $G \in \mathcal{A}(\boldsymbol{E}) \cup \mathcal{S}(\boldsymbol{E})$. Let $\boldsymbol{F} \subset \boldsymbol{E}$ be a linear subspace stable with respect to G, i.e. $G(\boldsymbol{F}) \subset \boldsymbol{F}$. Then, $\boldsymbol{F}^{\perp}$ is also stable with respect to G, i.e. $G\left(\boldsymbol{F}^{\perp}\right) \subset \boldsymbol{F}^{\perp}$.

PROOF.– Let $\mathbf{x} \in \boldsymbol{F}$ and $\mathbf{y} \in \boldsymbol{F}^{\perp}$. Since $\mathbf{y}$ belongs to $\boldsymbol{F}^{\perp}$, we know that by definition $\forall \mathbf{z} \in \boldsymbol{F}, \; b(\mathbf{z}, \mathbf{y}) = 0$. Let us now calculate:

$$b(\mathbf{x}, G(\mathbf{y})) = \pm b(G(\mathbf{x}), \mathbf{y}), \quad [11.11]$$

but since $\boldsymbol{F}$ is stable with respect to G, then $G(\mathbf{x})$ belongs to $\boldsymbol{F}$, and since $\mathbf{y}$ belongs to $\boldsymbol{F}^{\perp}$, then the dot product above is null. Therefore, if $\mathbf{y}$ belongs to $\boldsymbol{F}^{\perp}$, then $G(\mathbf{y})$ is orthogonal to all $\mathbf{x}$ of $\boldsymbol{F}$ as well, which proves that $G(\mathbf{y})$ belongs to $\boldsymbol{F}^{\perp}$, and thus the latter is stable with respect to G.

11.1.2. *Lack of diagonalization in pseudo-Euclidean spaces*

Let us recall the following definition:

DEFINITION 11.2.– An endomorphism u over $(\boldsymbol{E}, b)$ is called diagonalizable if and only if there exists a basis of $\boldsymbol{E}$ formed by eigenvectors of u.

Since the characterization of diagonalization by means of the kernel theorem does not mention Euclidean or pseudo-Euclidean structures, the usual theorems hold:

THEOREM 11.1.– The following statements are equivalent:

1) the endomorphism u is diagonalizable;

2) the characteristic polynomial of the endomorphism u splits into linear factors over $\mathbb{R}$;

3) the space $\boldsymbol{E}$ can be expressed as a direct sum of spaces $\boldsymbol{E}_{\lambda_i}$, where λ_i are the eigenvalues of u and $\mathbf{E}_{\lambda_i}$ is the associated eigenspace.

It is clear that two eigenvectors associated with distinct eigenvalues are pseudo-orthogonal, even in the pseudo-Euclidean case:

LEMMA 11.1.– Let $\alpha \neq \beta$ be two distinct eigenvalues of a symmetric endomorphism u. Then, we have $\boldsymbol{E}_\alpha \oplus^\perp \boldsymbol{E}_\beta$: in other words, $\boldsymbol{E}_\alpha$ and $\boldsymbol{E}_\beta$ are in a direct orthogonal sum.

PROOF.– Let $\mathbf{x} \in \boldsymbol{E}_\alpha \cap \boldsymbol{E}_\beta$. Then, $u(\mathbf{x}) = \alpha\mathbf{x}$ and $u(\mathbf{x}) = \beta\mathbf{x}$. Hence, we deduce that $(\alpha - \beta)\mathbf{x} = \mathbf{0}$ and thus $\mathbf{x} = \mathbf{0}$. Let now $\mathbf{x} \in \boldsymbol{E}_\alpha$ and $\mathbf{y} \in \boldsymbol{E}_\beta$. Then, we have:

$$b(u(\mathbf{x}), \mathbf{y}) = b(\mathbf{x}, \mathbf{u}(\mathbf{y})) \Rightarrow \alpha b(\mathbf{x}, \mathbf{y}) = \beta b(\mathbf{x}, \mathbf{y}) \Rightarrow b(\mathbf{x}, \mathbf{y}) = 0. \qquad [11.12]$$

The following proposition is a consequence of this reminder:

COROLLARY 11.1.– Let u be a symmetric endomorphism over pseudo-Euclidean $(\boldsymbol{E}, b)$. Then, it is diagonalizable if and only if it admits a pseudo-orthonormal basis of eigenvectors.

PROOF.– Suppose that u (symmetric) is diagonalizable. It is possible to construct a pseudo-orthogonal basis of each eigenspace $\boldsymbol{E}_{\lambda_i}$. A merge of these bases forms a basis of $(\boldsymbol{E}, b)$ since $\boldsymbol{E}$ is the (direct orthogonal) sum of these eigenspaces. This basis is of course pseudo-orthogonal. If we denote this basis by $\mathbf{e}_i, i \in [1, n]$, we have, via corollary 6.1, that $q(\mathbf{e}_i) \neq 0$ (since the form b is nondegenerate) and the basis can be made pseudo-orthonormal. The reciprocal is evident (by definition).

The lack of diagonalization, brutal for our following considerations, is now illustrated:

THEOREM 11.2.– Within pseudo-Euclidean spaces, there (always) exist symmetric endomorphisms that are not diagonalizable.

PROOF.– This lack of diagonalization of symmetric endomorphisms can be obtained systematically by referring to the existence of the isotropic cone. In fact, consider $\mathbf{f}_1$ such that $q(\mathbf{f}_1) = 0$, and $\mathbf{f}_2$ non-isotropic, orthogonal to $\mathbf{f}_1$. Then, consider the endomorphism g, defined as:

$$\forall \mathbf{x} \in \boldsymbol{E}, \;\; g(\mathbf{x}) = [b(\mathbf{f}_1, \mathbf{x})\,\mathbf{f}_2 - b(\mathbf{f}_2, \mathbf{x})\,\mathbf{f}_1]. \qquad [11.13]$$

It is evident that this endomorphism is skew-symmetric and thus that g^2 is symmetric (see proposition 11.2). Let us calculate:

$$g^2(\mathbf{x}) = b(\mathbf{f}_1, \mathbf{x})\,[b(\mathbf{f}_1, \mathbf{f}_2)\,\mathbf{f}_2 - q(\mathbf{f}_2)\,\mathbf{f}_1] - b(\mathbf{f}_2, \mathbf{x})\,[q(\mathbf{f}_1)\,\mathbf{f}_2 - b(\mathbf{f}_1, \mathbf{f}_2)\,\mathbf{f}_2], \quad [11.14]$$

by taking into account the different conditions regarding $\mathbf{f}_1, \mathbf{f}_2$, we deduce that:

$$g^2(\mathbf{x}) = -q(\mathbf{f}_2)\,b(\mathbf{f}_1, \mathbf{x})\,\mathbf{f}_1. \qquad [11.15]$$

Suppose that g^2 is diagonalizable. Then there exists a basis of eigenvectors for g^2. However, we have $\mathbf{E} \neq \langle \mathbf{f}_1 \rangle \oplus^{\perp} \langle \mathbf{f}_1 \rangle^{\perp}$. In particular, for whichever basis, there exists a vector $\mathbf{e}_n$ such that:

$$\mathbf{e}_n \notin \langle \mathbf{f}_1 \rangle + \langle \mathbf{f}_1 \rangle^{\perp}. \qquad [11.16]$$

Then, let us calculate:

$$g^2(\mathbf{e}_n) = -q(\mathbf{f}_2)\,b(\mathbf{f}_1, \mathbf{e}_n)\,\mathbf{f}_1, \qquad [11.17]$$

this vector cannot be an eigenvector of g^2 as in that case, we would have:

$$g^2(\mathbf{e}_n) = \lambda_n \mathbf{e}_n \Rightarrow \lambda_n \mathbf{e}_n = -q(\mathbf{f}_2)\,b(\mathbf{f}_1, \mathbf{e}_n)\,\mathbf{f}_1, \qquad [11.18]$$

and since $\mathbf{e}_n \notin \langle \mathbf{f}_1 \rangle + \langle \mathbf{f}_1 \rangle^{\perp}$, the vectors $\mathbf{e}_n$ and $\mathbf{f}_1$ are not collinear. So, the only remaining possibility is that $\lambda_n = -q(\mathbf{f}_2)\,b(\mathbf{f}_1, \mathbf{e}_n) = 0$. But then, this would imply that $b(\mathbf{f}_1, \mathbf{e}_n) = 0$ and therefore $\mathbf{e}_n \in \langle \mathbf{f}_1 \rangle^{\perp}$, which is excluded by construction.

11.1.3. *Isomorphism between skew-symmetric operators and* $\Lambda^2 E$ *elements*

Following a reasoning regarding the dimensions, it is rather straightforward to show that the space of skew-symmetric operators over $\boldsymbol{E}$ is isomorphic to the set $\Lambda^2 \boldsymbol{E}$. Here, we address a concrete method for constructing such an isomorphism:

THEOREM 11.3.– For all skew-symmetric endomorphism $g \in \mathcal{A}(\boldsymbol{E})$, there exists a unique element $\mathbf{G} \in \Lambda^2 \boldsymbol{E}$ such that:

$$\forall \mathbf{x} \in \boldsymbol{E}, g(\mathbf{x}) = \mathbf{G} : \mathbf{x}. \qquad [11.19]$$

Moreover, the map that associates $\mathbf{G}$ with g is an isomorphism of $\mathcal{A}(\boldsymbol{E})$ in $\Lambda^2 \boldsymbol{E}$.

PROOF.– Since g is skew-symmetric (with respect to the dot product), the map defined by $(\mathbf{x}, \mathbf{y}) \rightarrow b(g(\mathbf{x}), \mathbf{y})$ is an alternating bilinear form. From the construction theorem of the exterior algebra (see theorem 2.1), there exists a unique linear form l over $\Lambda^2 \boldsymbol{E}$ such that:

$$\forall \mathbf{x}, \mathbf{y} \in \boldsymbol{E}, \ \ l(\mathbf{x} \wedge \mathbf{y}) = b(g(\mathbf{x}), \mathbf{y}). \qquad [11.20]$$

By employing the Riesz theorem of representation, this linear form can be written in a unique way as $l(\mathbf{x} \wedge \mathbf{y}) = B_2(-\mathbf{G}, \mathbf{x} \wedge \mathbf{y}) = -\mathbf{G} : (\mathbf{x} \wedge \mathbf{y})$, and therefore we immediately have:

$$b(g(\mathbf{x}), \mathbf{y}) = B_2(-\mathbf{G}, \mathbf{x} \wedge \mathbf{y}) = B_2(\mathbf{G}, \mathbf{y} \wedge \mathbf{x}) = b(\mathbf{G} : \mathbf{x}, \mathbf{y}). \qquad [11.21]$$

Then, we deduce that:

$$\forall \mathbf{x} \in \boldsymbol{E}, \ \ g(\mathbf{x}) = \mathbf{G} : \mathbf{x}. \qquad [11.22]$$

The uniqueness of the element $\mathbf{G}$ enables the definition of a map $\gamma : \mathcal{A}(\boldsymbol{E}) \mapsto \Lambda^2 \boldsymbol{E}$. It is defined implicitly by the relation:

$$\forall g \in \mathcal{A}(\mathbf{E}), \forall \mathbf{x}, \mathbf{y} \in \boldsymbol{E}, \ \ b(g(\mathbf{x}), \mathbf{y}) = -\gamma(g) : \mathbf{x} \wedge \mathbf{y}. \qquad [11.23]$$

It is clear that this map γ is linear. We shall show directly (without prior arguments regarding the dimension) that it is bijective. Let $\mathbf{G} \in \Lambda^2 \boldsymbol{E}$. Then, the linear map $\mathbf{x} \mapsto \mathbf{G} : \mathbf{x}$ is skew-symmetric since:

$$b(\mathbf{G} : \mathbf{x}, \mathbf{y}) = B_2(\mathbf{G}, \mathbf{y} \wedge \mathbf{x}) = -B_2(\mathbf{G}, \mathbf{x} \wedge \mathbf{y}) = -b(\mathbf{G} : \mathbf{y}, \mathbf{x}). \qquad [11.24]$$

According to the previous result, we know that there exists a unique $\mathbf{H} \in \Lambda^2 \boldsymbol{E}$ such that:

$$\forall \mathbf{x} \in \boldsymbol{E}, \ \ \mathbf{G} : \mathbf{x} = \mathbf{H} : \mathbf{x} \Leftrightarrow \forall \mathbf{x} \in \boldsymbol{E} (\mathbf{G} - \mathbf{H}) : \mathbf{x} = \mathbf{0}. \qquad [11.25]$$

From this equality, we instantly deduce that we have $\mathbf{G} = \mathbf{H}$ since then:

$$\forall \mathbf{x}, \mathbf{y} \in E,\ [(\mathbf{G} - \mathbf{H}) : \mathbf{x}] : \mathbf{y} = 0 \Leftrightarrow \forall \mathbf{x}, \mathbf{y} \in E,\ (\mathbf{G} - \mathbf{H}) : (\mathbf{y} \wedge \mathbf{x}) = 0. \qquad [11.26]$$

This guarantees that $\mathbf{G} - \mathbf{H} = \mathbf{0}$ since $\mathbf{G} - \mathbf{H}$ can be decomposed in a pseudo-orthonormal basis of the form $\mathbf{e}_i \wedge \mathbf{e}_j$. Therefore, the surjectivity is established since we have in fact discovered that $\gamma(\mathbf{G} : (\cdot)) = \mathbf{G}$ for all $\mathbf{G}$ of $\Lambda^2 E$. Suppose now that $\gamma(g) = \mathbf{0}$. Then, this implies that:

$$\forall \mathbf{x}, \mathbf{y},\ \gamma(g) : (\mathbf{x} \wedge \mathbf{y}) = 0, \qquad [11.27]$$

now, we know, by definition of function γ, that:

$$\forall \mathbf{x}, \mathbf{y},\ \gamma(g) : (\mathbf{x} \wedge \mathbf{y}) = -b(g(\mathbf{x}), \mathbf{y}). \qquad [11.28]$$

In particular, if the right-hand side is null for all $\mathbf{x}, \mathbf{y}$, this implies that $g(\mathbf{x})$ is null for all $\mathbf{x}$. Therefore, the injectivity is established.

11.2. Decomposability within $\Lambda^2 E$

11.2.1. *Simple decomposability*

DEFINITION 11.3.– An element $\mathbf{E} \in \Lambda^2 E$ is called:

1) decomposable if there exist $\mathbf{x}, \mathbf{y} \in E$ such that $\mathbf{E} = \mathbf{x} \wedge \mathbf{y}$;

2) pseudo-orthodecomposable if there exist $\mathbf{x}, \mathbf{y}$ such that $b(\mathbf{x}, \mathbf{y}) = 0$ and $\mathbf{E} = \mathbf{x} \wedge \mathbf{y}$.

It is rather clear that any 2-vector that is decomposable is also orthodecomposable: if we denote by F the space generated by $\mathbf{x}, \mathbf{y}$, it is always possible to diagonalize the form b_F within F, obtaining a pseudo-orthogonal basis $\mathbf{e}_1, \mathbf{e}_2$ of F which allows us to write $\mathbf{x} \wedge \mathbf{y} = \alpha \mathbf{e}_1 \wedge \mathbf{e}_2$.

THEOREM 11.4.– Let $\mathbf{E}$ be a vector of $\Lambda^2 E$. Then, it is decomposable if and only if it satisfies $\mathbf{E} \wedge \mathbf{E} = \mathbf{0}$.

PROOF.– The direct part is straightforward: if $\mathbf{E} = \mathbf{f}_1 \wedge \mathbf{f}_2$, it is immediate that $\mathbf{E} \wedge \mathbf{E} = \mathbf{0}$.

As for the reciprocal statement, it can be shown via an inductive demonstration. Let us denote by $\mathcal{P}(k)$ the statement below, which depends on an integer k:

$$\mathcal{P}(k): \ \dim(\boldsymbol{E}) = k \Rightarrow$$
$$\left(\forall \mathbf{E} \in \Lambda^2 \boldsymbol{E}, [\mathbf{E} \wedge \mathbf{E} = \mathbf{0} \Rightarrow \ \mathbf{E} \text{ is decomposable}]\right). \quad [11.29]$$

We already know that $\mathcal{P}(2)$ and $\mathcal{P}(3)$ are true. In fact, when $\boldsymbol{E}$ is a space with a dimension $n \in \{2, 3\}$, then we know that all the elements of $\Lambda^2 \boldsymbol{E}$ necessarily satisfy $\mathbf{E} \wedge \mathbf{E} = \mathbf{0}$ and they are also decomposable: for $n = 2$, we work in $\Lambda^n \boldsymbol{E}$, and for $n = 3$, we work in $\Lambda^{n-1} \boldsymbol{E}$ (see proposition 8.7).

Suppose now, with a fixed $n \geq 4$, that $\mathcal{P}(k)$ is true for all $k \leq n - 1$. Let us show then that property $\mathcal{P}(n)$ is true. Let $\boldsymbol{E}$ be a space with a dimension n and $\mathbf{E} \in \Lambda^2 \boldsymbol{E}$. Let $\mathbf{e}_1, \cdots, \mathbf{e}_n$ be a pseudo-orthonormal basis of $\boldsymbol{E}$. *A priori*, we can write $\mathbf{E}$ in the following form:

$$\mathbf{E} = \sum_{1 \leq i < j < n} \alpha_{ij} \mathbf{e}_i \wedge \mathbf{e}_j + \left(\sum_{i=1}^{n-1} \alpha_{in} \mathbf{e}_i \right) \wedge \mathbf{e}_n := \mathbf{U} + \mathbf{u} \wedge \mathbf{e}_n. \quad [11.30]$$

Let us consider the contraction product of $\mathbf{E} \wedge \mathbf{E}$ with the vector $\mathbf{e}_n$. Since $\mathbf{U} \wedge \mathbf{U}$ can be developed with respect to some 4-blades that include only vectors $\mathbf{e}_1, \cdots, \mathbf{e}_{n-1}$, which are all orthogonal to $\mathbf{e}_n$, then, by proposition 8.2, we instantly see that:

$$\mathbf{U} \wedge \mathbf{U} : \mathbf{e}_n = \mathbf{0}, \quad [11.31]$$

and therefore we have:

$$\mathbf{U} \wedge \mathbf{u} \wedge \mathbf{e}_n : \mathbf{e}_n = \mathbf{0}. \quad [11.32]$$

Again, let us expand $\mathbf{U}$, as well as $\mathbf{u}$, with respect to the basis $\mathbf{e}_i \wedge \mathbf{e}_j$, $i < j < n$, and by employing proposition 8.2 and the pseudo-orthonormality of the basis $\mathbf{e}_i, i \in [1, n]$, we obtain:

$$q(\mathbf{e}_n) \mathbf{U} \wedge \mathbf{u} = \mathbf{0} \Rightarrow \mathbf{U} \wedge \mathbf{u} = \mathbf{0}, \quad [11.33]$$

consequently, we recover $\mathbf{U} \wedge \mathbf{u} \wedge \mathbf{e}_n = \mathbf{0}$ and $\mathbf{U} \wedge \mathbf{U} = \mathbf{0}$. Now, $\mathbf{U}$ is expressed with respect to $\Lambda^2 \boldsymbol{F}$, with $\boldsymbol{F}$ being the space generated by $\mathbf{e}_1, \cdots, \mathbf{e}_{n-1}$, which thus has a dimension lower than or equal to $n-1$. Since it satisfies $\mathbf{U} \wedge \mathbf{U} = \mathbf{0}$, it is decomposable in accordance with the induction hypothesis. Let us express it in the following form:

$$\mathbf{U} = \theta \mathbf{f}_1 \wedge \mathbf{f}_2 \quad [11.34]$$

with $\mathbf{f}_1, \mathbf{f}_2$ orthogonal and belonging to the space $\boldsymbol{F}$ generated by $\mathbf{e}_1, \cdots, \mathbf{e}_{n-1}$. Therefore, we have:

$$\mathbf{E} = \theta \mathbf{f}_1 \wedge \mathbf{f}_2 + \mathbf{u} \wedge \mathbf{e}_n. \quad [11.35]$$

We have seen that $\mathbf{U} \wedge \mathbf{u} = \mathbf{0}$, which implies that:

$$\mathbf{f}_1 \wedge \mathbf{f}_2 \wedge \mathbf{u} = \mathbf{0}, \quad [11.36]$$

since $\mathbf{f}_1, \mathbf{f}_2$ forms an independent set, this means that $\mathbf{u}$ is a linear combination of these elements. We can write $\mathbf{u} = u_1 \mathbf{f}_1 + u_2 \mathbf{f}_2$. Now, it remains that:

$$\mathbf{E} = \mathbf{f}_1 \wedge \mathbf{f}_2 + (u_1 \mathbf{f}_1 + u_2 \mathbf{f}_2) \wedge \mathbf{e}_n, \quad [11.37]$$

which shows that $\mathbf{E}$ is expressed within $\Lambda^2 \boldsymbol{H}$, with $\boldsymbol{H}$ being the space with a dimension three constructed over $\mathbf{f}_1, \mathbf{f}_2, \mathbf{e}_n$. Consequently, $\mathbf{E}$ is decomposable according to the inductive hypothesis.

REMARK 11.1.– Actually, a pseudo-Euclidean structure (and a contraction product) are not required for reaching the same conclusion (although, in the end, all finite-dimensional linear spaces are isomorphic to $\mathbb{R}^n$, and since it is always possible to equip $\mathbb{R}^n$ with a pseudo-dot product, $\Lambda^2 \boldsymbol{E}$ can be always equipped with a pseudo-Euclidean structure and thus with a contraction product). The advantage of reasoning with the contraction product is that it is quite natural to write. ■

The question now arising is the following:

QUESTION 11.2.– What happens if a vector of $\Lambda^2 \boldsymbol{E}$ is not decomposable (i.e. it is not a 2-blade)? What can be said regarding its structure?

11.2.2. *p-Decomposability*

In order to develop an answer to this question, we need to follow the previous decomposition. Let us introduce the following definitions:

DEFINITION 11.4.– Let $\mathbf{E}$ be an element of $\Lambda^2 \boldsymbol{E}$ and $p \geq 1$. We say that this 2-vector is:

1) p-(pseudo-ortho-) decomposable when there exists $\mathbf{f}_1, \cdots, \mathbf{f}_{2p}$ a (pseudo-orthogonal) family of $\boldsymbol{E}$ as well as a family $\eta_1, \cdots, \eta_p$ of reals such that:

$$\mathbf{E} = \sum_{i=1}^{i=p} \eta_i \mathbf{f}_{2i-1} \wedge \mathbf{f}_{2i}. \qquad [11.38]$$

2) Exactly p-(pseudo-ortho-) decomposable if it is p-(pseudo-ortho-) decomposable and not $(p-1)$-(pseudo-ortho-) decomposable.

There is a rather convenient algebraic characterization of p-decomposability:

THEOREM 11.5.– Let $\mathbf{E}$ be an element of $\Lambda^2 \boldsymbol{E}$. Then, it is p-decomposable if and only if it satisfies $\mathbf{E}\wedge^{p+1} = \mathbf{0}$.

Before beginning the demonstration of this theorem, let us draw an immediate corollary from it:

COROLLARY 11.2.– Let $\mathbf{E} \in \Lambda^2 \boldsymbol{E}$ be a p-decomposable 2-vector. Then, it is exactly p-decomposable if and only if it also satisfies $\mathbf{E}\wedge^{p} \neq \mathbf{0}$.

PROOF.– The condition is obviously necessary. In fact, let us suppose that:

$$\mathbf{E} = \sum_{i=1}^{i=p} \theta_i \mathbf{f}_{2i-1} \wedge \mathbf{f}_{2i}. \qquad [11.39]$$

Then, if we expand $\mathbf{E}\wedge^{p+1}$, the exterior product of $2p+2$ terms appears in each term. The space generated by the $\mathbf{f}_i$ is a space with a dimension $2p$, and consequently each term includes exterior products over dependent families, and therefore each term is null. Consider now the reciprocal statement, which can be shown by induction over the dimension of space $\boldsymbol{E}$. By fixing n, the dimension of $\boldsymbol{E}$, let us denote by $\mathcal{P}(\cdot)$ the following statement, which depends on integer n:

$$\mathcal{P}(n): \ \forall \mathbf{E} \in \Lambda^2 \boldsymbol{E}, \ \left[\forall p \ \left[\mathbf{E}\wedge^{p+1} = \mathbf{0} \ \Rightarrow \mathbf{E} \text{ is } p\text{-decomposable}\right]\right]. \qquad [11.40]$$

For $n = 2$ and $n = 3$, whichever is $p \geq 1$, we always have the fact that $\mathbf{E}\wedge^{p+1} = \mathbf{0}$, and also the fact that any vector of $\Lambda^2 \boldsymbol{E}$ is p-decomposable since it is 2-decomposable (see the previous theorem 11.4). Therefore, $\mathcal{P}(2)$ and $\mathcal{P}(3)$ are true. Suppose that

$\mathcal{P}(k)$ is true for all integers $k \leq n-1$ (with $n > 3$) and let us show that this entails $\mathcal{P}(n)$. Consider $\mathbf{E} \in \Lambda^2 E$ such that $\mathbf{E}\wedge^{p+1} = \mathbf{0}$. Let us expand $\mathbf{E}$ over a pseudo-orthonormal basis $\mathbf{e}_1, \cdots, \mathbf{e}_n$:

$$\mathbf{E} = \sum_{1\leq i<j<n} \alpha_{ij}\mathbf{e}_i \wedge \mathbf{e}_j + \sum_{j=1}^{j=n} \alpha_{jn}\mathbf{e}_j \wedge \mathbf{e}_n = \mathbf{U} + \mathbf{u} \wedge \mathbf{e}_n, \quad [11.41]$$

it can be easily seen that:

$$\mathbf{E}\wedge^{p+1} = \mathbf{U}\wedge^{p+1} + (p+1)\,\mathbf{U}\wedge^{p} \wedge\mathbf{u} \wedge \mathbf{e}_n, \quad [11.42]$$

and we deduce:

$$\mathbf{E}\wedge^{p+1} : \mathbf{e}_n = \mathbf{0} \Leftrightarrow \mathbf{U}\wedge^{p+1} : \mathbf{e}_n + \mathbf{U}\wedge^{p} \wedge\mathbf{u} \wedge \mathbf{e}_n : \mathbf{e}_n = \mathbf{0}, \quad [11.43]$$

now, the term $\mathbf{U}\wedge^{p+1}$ can be developed by introducing $2(p+1)$-blades written with respect to $\mathbf{e}_1, \cdots, \mathbf{e}_{n-1}$. According to proposition 8.2, we then have $\mathbf{U}\wedge^{p+1} : \mathbf{e}_n = \mathbf{0}$, and hence we deduce $\mathbf{U}\wedge^{p} \mathbf{u} \wedge \mathbf{e}_n : \mathbf{e}_n$. Once again, by expanding $\mathbf{U}$ over the basis of $\mathbf{e}_1, \cdots, \mathbf{e}_{n-1}$, together with the vector $\mathbf{u}$, we immediately have, via proposition 8.2:

$$\mathbf{U}\wedge^{p} \mathbf{u} \wedge \mathbf{e}_n : \mathbf{e}_n = \mathbf{0} \Leftrightarrow q(\mathbf{e}_n)\,\mathbf{U}\wedge^{p} \mathbf{u} = \mathbf{0} \Leftrightarrow \mathbf{U}\wedge^{p} \mathbf{u} = \mathbf{0}, \quad [11.44]$$

since we actually have $q(\mathbf{e}_n) = \pm 1 \neq 0$. By considering again the equality $\mathbf{E}\wedge^{p+1} = \mathbf{0}$, we can deduce that:

$$\mathbf{U}\wedge^{p+1} = \mathbf{0} \;\; and \;\; (\mathbf{U}\wedge^{p}) \wedge \mathbf{u} = \mathbf{0}. \quad [11.45]$$

In virtue of the inductive hypothesis regarding the dimension of $\boldsymbol{E}$, since $\mathbf{U}$ is constructed within $\Lambda^2 \boldsymbol{F}$, with $\boldsymbol{F}$ the space with a dimension $n-1$ constructed over $\mathbf{e}_1, \cdots, \mathbf{e}_{n-1}$, it is possible to write that:

$$\mathbf{U}\wedge^{p+1} = \mathbf{0} \Rightarrow \mathbf{U} = \sum_{i=1}^{i=p} \theta_i \mathbf{f}_{2i-1} \wedge \mathbf{f}_{2i} \quad [11.46]$$

with $\mathbf{f}_1, \cdots, \mathbf{f}_{2p}$ belonging to the space $\boldsymbol{F}$. Then, it can be easily verified that:

$$\mathbf{U}\wedge^{p} = \left(\pi_{k=1}^{k=p}\theta_k\right) \mathbf{f}_1 \wedge \cdots \wedge \mathbf{f}_{2p}. \quad [11.47]$$

Suppose, at first, that $\mathbf{U}\wedge^p \neq 0$, in other words that $\forall k \in [1,p], \theta_k \neq 0$ and that $\mathbf{f}_1, \cdots, \mathbf{f}_{2p}$ is an independent family of $\boldsymbol{F}$. Then, in this case, since $\mathbf{e}_n$ is orthogonal to $\boldsymbol{F}$ and $\mathbf{e}_n$ is orthogonal to $\mathbf{u}$, we have:

$$\mathbf{U} \wedge^p \wedge \mathbf{u} \wedge \mathbf{e}_n = \mathbf{0} \Rightarrow \mathbf{U} \wedge^p \mathbf{u} = \mathbf{0} \quad \Rightarrow \mathbf{u} = \sum_{k=1}^{k=2p} u_i \mathbf{f}_i, \tag{11.48}$$

since family $\mathbf{f}_1, \cdots, \mathbf{f}_{2p}$ is linearly independent but family $\mathbf{f}_1, \cdots, \mathbf{f}_{2p}, \mathbf{u}$ is linearly dependent. We can rewrite:

$$\mathbf{E} = \sum_{i=1}^{i=p} \left[u_{2i-1}\mathbf{f}_{2i-1} \wedge \mathbf{e}_n + u_{2i}\mathbf{f}_{2i} \wedge \mathbf{e}_n + \theta_i \mathbf{f}_{2i-1} \wedge \mathbf{f}_{2i} \right]. \tag{11.49}$$

Now, each element inside the bracket is a vector of $\Lambda^2 \boldsymbol{G}_i$, with $\boldsymbol{G}_i$ a linear space with a dimension three constructed over $\mathbf{f}_{2i-1}, \mathbf{f}_{2i}, \mathbf{e}_n$. Therefore, it is decomposable and it is possible to write:

$$\mathbf{E} = \sum_{i=1}^{i=p} \alpha_i \mathbf{g}_{2i-1} \wedge \mathbf{g}_{2i}. \tag{11.50}$$

Now, if we have $\mathbf{U}\wedge^p = \mathbf{0}$, since it is an element of $\Lambda^2 \boldsymbol{F}$ with $\boldsymbol{F}$ having a dimension $n-1$, by the hypothesis of induction, we know that $\mathbf{U}$ is $(p-1)$-decomposable. Since we had:

$$\mathbf{E} = \mathbf{U} + \mathbf{u} \wedge \mathbf{e}_n, \tag{11.51}$$

we immediately deduce that $\mathbf{E}$ is indeed p-decomposable.

A rather important corollary stems from this fundamental decomposition theorem:

COROLLARY 11.3.– Let $\boldsymbol{E}$ be a space with a dimension n and $\left\lfloor \frac{n}{2} \right\rfloor$ be the floor function of $\frac{n}{2}$. Then, any vector of $\Lambda^2 \boldsymbol{E}$ is $\left\lfloor \frac{n}{2} \right\rfloor$-decomposable.

PROOF.– Let $\mathbf{E}$ be an element of $\Lambda^2 \boldsymbol{E}$. In general, it can be written in the following form:

$$\mathbf{E} = \sum_{i<j} \alpha_{ij} \mathbf{e}_i \wedge \mathbf{e}_j \tag{11.52}$$

with $\mathbf{e}_i, i \in [1,n]$ an orthonormal basis of $\boldsymbol{E}$. Let us denote by $p = \left\lfloor \frac{n}{2} \right\rfloor$. Then, in the expansion of $\mathbf{E}\wedge^{p+1}$, only exterior products with $2p+2$ vectors of $\boldsymbol{E}$ are present.

Because of $2p + 2 > n$, some vectors of $\boldsymbol{E}$ within each exterior product strictly exceed the dimension of $\boldsymbol{E}$. The vectors are thus necessarily linearly dependent and the corresponding exterior products are necessarily null. Then, since $\mathbf{E}\wedge^{p+1} = 0$, by the previous theorem, we can see that $\mathbf{E}$ is indeed $\left\lfloor \frac{n}{2} \right\rfloor$- decomposable.

THEOREM 11.6 (Cartan's lemma).– Let $\boldsymbol{E}$ be a space with a dimension n, let p be an integer $p \leq n$ and $\mathbf{x}_1, \cdots, \mathbf{x}_p$ be an independent family of $\boldsymbol{E}$. Then, we have:

$$\forall \mathbf{y}_1, \cdots, \mathbf{y}_p, \ \sum_{i=1}^{i=p} \mathbf{x}_i \wedge \mathbf{y}_i = \mathbf{0} \Leftrightarrow$$

$$\exists \mathbb{A} \textit{ a symmetric matrix}, \ \forall i \in [1,p], \ \mathbf{y}_i = \sum_{j=1}^{j=p} A_{ij} \mathbf{x}_j. \qquad [11.53]$$

PROOF.– For all j of $[1,p]$, consider the $p-1$-blade $(\mathbf{x}_1 \wedge \cdots \mathbf{x}_p / [\mathbf{x}_j])$. Then, we have:

$$(\mathbf{x}_1 \wedge \cdots \mathbf{x}_p / [\mathbf{x}_j]) \wedge \left(\sum_{i=1}^{i=p} \mathbf{x}_i \wedge \mathbf{y}_i \right) = \mathbf{0}. \qquad [11.54]$$

By expanding this exterior product, we immediately see that we have:

$$\forall j \in [1,p], (-1)^{p+j} \mathbf{x}_1 \wedge \cdots \wedge \mathbf{x}_p \wedge \mathbf{y}_j = \mathbf{0}. \qquad [11.55]$$

Since family $\mathbf{x}_1, \cdots, \mathbf{x}_p$ is linearly independent, this proves that each vector $\mathbf{y}_j$ is a linear combination of the $\mathbf{x}_i$. Thus, we can actually write that we have:

$$\forall i \in [1,p], \ \mathbf{y}_i = \sum_{j=1}^{j=p} A_{ij} \mathbf{x}_j. \qquad [11.56]$$

It remains to demonstrate that the matrix of A_{ij} is symmetric. For this purpose, let us consider:

$$\left(\sum_{k=1}^{k=p} \mathbf{x}_k \wedge \mathbf{y}_k \right) \wedge (\mathbf{x}_1 \wedge \cdots \wedge \mathbf{x}_p / [\mathbf{x}_i, \mathbf{x}_j]) = \mathbf{0} \qquad [11.57]$$

by developing, we then obtain:

$$\forall i \neq j, \ \mathbf{x}_i \wedge \mathbf{y}_i \wedge (\mathbf{x}_1 \wedge \cdots \wedge \mathbf{x}_p / [\mathbf{x}_i, \mathbf{x}_j]) + \mathbf{x}_j \wedge \mathbf{y}_j \wedge (\mathbf{x}_1 \wedge \cdots \wedge \mathbf{x}_p / [\mathbf{x}_i, \mathbf{x}_j]) = \mathbf{0}. \qquad [11.58]$$

Next, we develop $\mathbf{y}_i$ and $\mathbf{y}_j$ with respect to the basis of $\mathbf{x}_k$, obtaining:

$$\forall i \neq j, \ \ \mathbf{x}_i \wedge (A_{ij}\mathbf{x}_j) \wedge (\mathbf{x}_1 \wedge \cdots \wedge \mathbf{x}_p / [\mathbf{x}_i, \mathbf{x}_j]) + \mathbf{x}_j \wedge (A_{ji}\mathbf{x}_i) \wedge (\mathbf{x}_1 \wedge \cdots \wedge \mathbf{x}_p / [\mathbf{x}_i, \mathbf{x}_j]) = \mathbf{0}. \quad [11.59]$$

Now, it is immediate that we have:

$$\mathbf{x}_i \wedge \mathbf{x}_j \wedge (\mathbf{x}_1 \wedge \cdots \wedge \mathbf{x}_p / [\mathbf{x}_i, \mathbf{x}_j]) = -\mathbf{x}_j \wedge \mathbf{x}_i \wedge (\mathbf{x}_1 \wedge \cdots \wedge \mathbf{x}_p / [\mathbf{x}_i, \mathbf{x}_j]) \neq \mathbf{0}, \quad [11.60]$$

(since family $\mathbf{x}_1, \cdots, \mathbf{x}_p$ is linearly independent). Hence, we deduce that:

$$\forall i \neq j \in [1, p], \ \ A_{ij} = A_{ji}, \quad [11.61]$$

and the matrix is indeed symmetric. For the reciprocal statement, the direct calculation can be quickly performed:

$$\sum_{k=1}^{k=p} \mathbf{x}_k \wedge \mathbf{y}_k = \sum_{k=1}^{k=p} \sum_{j=1}^{j=p} A_{kj} \mathbf{x}_k \wedge \mathbf{x}_j, \quad [11.62]$$

by changing notation $k, j \leftrightarrow j, k$ and by employing simultaneously the symmetry of the matrix $\mathbb{A}$ and the skew symmetry of the exterior product, we have:

$$\sum_{k=1}^{k=p} \mathbf{x}_k \wedge \mathbf{y}_k = -\sum_{k=1}^{k=p} \mathbf{x}_k \wedge \mathbf{y}_k, \quad [11.63]$$

which of course implies the nullity of the summation.

11.2.3. *Interpretation in terms of operators*

The previous result should be interpreted as a pseudo-skew-symmetric operator. Before illustrating this point, let us demonstrate a useful lemma:

LEMMA 11.2.– Let $\mathbf{f}_1, \cdots, \mathbf{f}_r$ be vectors of $\boldsymbol{E}$ and let $\phi_1, \cdots, \phi_r$ be linear forms over $\boldsymbol{E}$. Then:

$$\boldsymbol{F} := \left\{ \sum_{i=1}^{i=r} \phi_i(\mathbf{x}) \, \mathbf{f}_i, \ \mathbf{x} \in \boldsymbol{E} \right\} \quad [11.64]$$

is a subspace with a dimension r if and only if the families $\mathbf{f}_i, i \in [1, r]$ and $\phi_i, i \in [1, r]$ are linearly independent, respectively in $\boldsymbol{E}$ and $\boldsymbol{E}^*$.

PROOF.– Suppose that the family of $\mathbf{f}_i$ is linearly dependent. Then, it is clear that $\boldsymbol{F}$ admits a generating family with a cardinality $r-1$ and therefore it cannot have a dimension r. Now, if the ϕ_i are linearly dependent, we can suppose, up to a permutation of indexes, that $\phi_r = \sum_{i=1}^{r-1} \alpha_i \phi_i$. We thus have:

$$\forall \mathbf{x}, \ \sum_{i=1}^{i=r} \phi_i(\mathbf{x}) \mathbf{f}_i = \sum_{i=1}^{r-1} \phi_i(\mathbf{x}) (\mathbf{f}_i + \alpha_i \mathbf{f}_r), \qquad [11.65]$$

once again, the generating family of the space $\boldsymbol{F}$ has a cardinality equal to $r-1$ and thus this space cannot have a dimension r. Therefore, by contraposition: $\boldsymbol{F}$ with a dimension r implies that the $\mathbf{f}_i, \phi_i$ are linearly independent families. Let us suppose now that each of them is a linearly independent family. Suppose that:

$$\sum_i \phi_i(\mathbf{x}) \mathbf{f}_i = 0. \qquad [11.66]$$

Since the family of the $\mathbf{f}_i$ is linearly independent, this implies that $\forall i \in [1, r], \ \phi_i(\mathbf{x}) = 0$. Since the family of the ϕ_i is linearly independent, this implies that $\mathbf{x}$ belongs to a linear subspace with a dimension $n-r$. Therefore, the kernel of the map $\mathbf{x} \mapsto \sum_{i=1}^{i=r} \phi_i(\mathbf{x}) \mathbf{f}_i$ has a dimension $n-r$. Then, according to the rank formula, its image has a dimension r.

PROPOSITION 11.5.– Let $g := \gamma^{-1}(\mathbf{E})$ be a pseudo-skew-symmetric endomorphism associated with $\mathbf{E}$. Then, $\mathbf{E}$ is exactly p-decomposable if and only if the image of g has a dimension $2p$.

PROOF.– Suppose that $\mathbf{E}$ is exactly p-decomposable. Then, according to definition 11.4 and corollary 11.2, we simultaneously have:

$$\mathbf{E} = \sum_{i=1}^{i=p} \eta_i \mathbf{f}_{2i-1} \wedge \mathbf{f}_{2i}, \ \ \textit{and}\ \mathbf{E}\wedge^p \neq \mathbf{0}, \qquad [11.67]$$

in particular, the latter condition entails that:

$$\left(\pi_{i=1}^{i=p} \eta_i\right) \mathbf{f}_1 \wedge \cdots \wedge \mathbf{f}_{2p} \neq \mathbf{0}, \qquad [11.68]$$

therefore, family $\mathbf{f}_1, \cdots, \mathbf{f}_{2p}$ is necessarily a linearly independent family and all the η_i are non-null. We then have:

$$\mathbf{E} : \mathbf{x} = \sum_{i=1}^{i=p} \eta_i \left[b(\mathbf{f}_{2i}, \mathbf{x}) \mathbf{f}_{2i-1} - b(\mathbf{f}_{2i-1}, \mathbf{x}) \mathbf{f}_{2i} \right], \qquad [11.69]$$

if we set $\mathbf{g}_{2i-1} = -\eta_i \mathbf{f}_{2i}$, $\mathbf{g}_{2i} = \eta_i \mathbf{f}_{2i-1}$, since none of the η_i is null, this implies that the family of the $\mathbf{g}_j, j \in [1, 2p]$ is linearly independent. We thus have:

$$Im\left(\gamma^{-1}(\mathbf{E})\right) = \left\{\sum_{i=1}^{i=p} b(\mathbf{f}_{2i-1}, \mathbf{x})\, \mathbf{g}_{2i-1} + b(\mathbf{f}_{2i}, \mathbf{x})\, \mathbf{g}_{2i}, \mathbf{x} \in \boldsymbol{E}\right\}. \qquad [11.70]$$

Now, since b is nondegenerate and since the $\mathbf{f}_k, k \in [1, 2p]$ are linearly independent, the family of $b(\mathbf{f}_k, \cdot)$ is linearly independent in $\boldsymbol{E}^*$. According to lemma 11.2, we therefore deduce that $\gamma^{-1}(\mathbf{E})$ has indeed an image with a dimension $2p$. For the reciprocal statement, we employ the fact that since any vector of $\Lambda^2 \boldsymbol{E}$ is at least $\lfloor \frac{n}{2} \rfloor$ decomposable, if it is non-null there exists a unique integer $q \in \left[1, \lfloor \frac{n}{2} \rfloor\right]$ such that:

$$\mathbf{E}\wedge^{q+1} = \mathbf{0}, \quad \mathbf{E}\wedge^{q} \neq \mathbf{0}. \qquad [11.71]$$

In particular, in this case, the image of $\gamma^{-1}(\mathbf{E})$ has a dimension $2q$. Since we know that, by hypothesis, its dimension is equal to $2p$, this means that $p = q$ and therefore we can see that $\mathbf{E}$ is p-decomposable.

11.2.4. *Pseudo-orthodecomposability*

An important issue concerning p-decomposability (with $p \geq 2$) of elements of $\Lambda^2 \boldsymbol{E}$ consists of knowing whether this decomposition can additionally become pseudo-orthogonal. A necessary condition is shown in the following theorem:

THEOREM 11.7.– Let $\mathbf{E} \in \Lambda^2 \boldsymbol{E}$ be a 2-vector in a space $(\boldsymbol{E}, b)$. Then, if it is pseudo-orthodecomposable, the (symmetric) endomorphism g^2 (such that $g := \gamma^{-1}(\mathbf{E})$) ***restricted*** to $Im(g)$ is diagonalizable.

REMARK 11.2.– The restriction of g^2 to $Im(g)$ is essential. In particular, as shown by the counter-example of theorem 11.2, an element of $\Lambda^2 \boldsymbol{E}$ can be pseudo-orthodecomposable without the endomorphism g^2 being diagonalizable over the whole $\boldsymbol{E}$. ■

PROOF.– Suppose that $\mathbf{E}$ is exactly p-pseudo-orthodecomposable and it can be written in the following form:

$$\mathbf{E} = \sum_{i=1}^{i=p} \eta_i \mathbf{f}_{2i-1} \wedge \mathbf{f}_{2i} \qquad [11.72]$$

with $\mathbf{f}_k$ forming a pseudo-orthogonal independent family. By employing proposition 8.2 and knowing that $g(\mathbf{x}) = \mathbf{E} : \mathbf{x}$, we can easily calculate:

$$g(\mathbf{x}) = \sum_{i=1}^{i=p} \eta_i \left[b(\mathbf{f}_{2i}, \mathbf{x}) \mathbf{f}_{2i-1} - b(\mathbf{f}_{2i-1}, \mathbf{x}) \mathbf{f}_{2i} \right]. \qquad [11.73]$$

By taking into account the orthogonality two by two of the $\mathbf{f}_k$ and by using once more proposition 8.2, we calculate again without any effort:

$$g^2(\mathbf{x}) = -\sum_{i=1}^{i=p} \eta_i^2 \left[b(\mathbf{f}_{2i-1}, \mathbf{x}) q(\mathbf{f}_{2i}) \mathbf{f}_{2i-1} + b(\mathbf{f}_{2i}, \mathbf{x}) q(\mathbf{f}_{2i-1}) \mathbf{f}_{2i} \right]. \qquad [11.74]$$

In particular, we can deduce that:

$$g^2(\mathbf{f}_{2j-1}) = -\eta_j^2 q(\mathbf{f}_{2j}) q(\mathbf{f}_{2j-1}) \mathbf{f}_{2j-1}, \quad g^2(\mathbf{f}_{2j}) = -\eta_j^2 q(\mathbf{f}_{2j}) q(\mathbf{f}_{2j-1}) \mathbf{f}_{2j}. \qquad [11.75]$$

Therefore, the basis of $\mathbf{f}_k$ is indeed an eigenbasis of g^2 when it is restricted to $Im(g)$.

We shall see that in the Euclidean case that the reverse is always true:

THEOREM 11.8 (reciprocal, euclidean case).– Let $(E, \cdot)$ be a ***Euclidean*** space and $\mathbf{E}$ an element of $\Lambda^2 E$. Suppose that the endomorphism $g := \gamma^{-1}(\mathbf{E})$ is such that g^2 restricted to $(Im(g), \cdot)$ is diagonalizable. Then, $\mathbf{E}$ is orthodecomposable.

Before continuing with the demonstration, let us immediately illustrate a very important corollary:

COROLLARY 11.4.– Let $(E, \cdot)$ be a Euclidean space. Then, any element of $\Lambda^2 E$ is exactly p-orthodecomposable for a certain integer p.

Actually, the space $(Im(g), \cdot)$ remains Euclidean and g^2 restricted to $Im(g)$ remains symmetric. In the Euclidean case, all symmetric endomorphisms are diagonalizable, which then concludes the justification of the corollary.

PROOF.– For greater clarity, we shall denote by G the restriction of g to the subspace $(Im(g), \cdot)$.

1) In a first phase, it is clear that G is an isomorphism since:

$$\mathbf{x} \in \ker G \Leftrightarrow \{ g(\mathbf{x}) = \mathbf{0} \text{ and } \mathbf{x} \in Im(g) \} \Leftrightarrow \mathbf{x} \in \ker(g) \bigcap Im(g). \qquad [11.76]$$

Since g is symmetric and $(\boldsymbol{E}, \cdot)$ is Euclidean (therefore pseudo-Euclidean), we know via proposition 11.3 that $Im\,(g) = \ker\,(g)^{\perp}$. Now, within a Euclidean space the subspaces are never degenerate and thus we indeed have $Im\,(g) \bigcap Im\,(g)^{\perp} = \{\mathbf{0}\}$ (see corollary 6.3), hence $\mathbf{x} = \mathbf{0}$ and the fact that G is invertible.

2) Consequently, besides being diagonalizable, G^2 is also invertible. Since G^2 is diagonalizable, let $\mathbf{e}_1$ be an eigenvector of G^2, with λ_1 being the associated eigenvalue. We set $\mathbf{e}_2 = G\,(\mathbf{e}_1)$. Due to the skew symmetry of G, it is straightforward that $\mathbf{e}_1$ and $\mathbf{e}_2$ are orthogonal. Furthermore, $\mathbf{e}_2$ satisfies $G^2\,(\mathbf{e}_2) = G^3\,(\mathbf{e}_1) = G\,(\lambda_1 \mathbf{e}_1) = \lambda_1 \mathbf{e}_2$. Finally, it is clear that $\mathbf{e}_2 \neq 0$, since $G\,(\mathbf{e}_2) = \lambda_1 \mathbf{e}_1 \neq \mathbf{0}$, which is due to the fact that $\mathbf{e}_1$ is an eigenvector (thus non-null) and $\lambda_1 \neq 0$ since G^2 is invertible.

3) As a result, the space $\boldsymbol{F}_1 := \langle \mathbf{e}_1, \mathbf{e}_2 \rangle$ is a space with a dimension two, stable with respect to G, since $G\,(\mathbf{e}_1) = \mathbf{e}_2$ and $G\,(\mathbf{e}_2) = \lambda_1 \mathbf{e}_1$. Now we set G_1 to be the restriction of G to $\boldsymbol{F}_1$. According to the above calculations, it is immediate that the restriction of G to $\boldsymbol{F}_1$ is proportional to $\lambda_1 \mathbb{I}_1$, where $\mathbb{I}_1$ designates the restriction of the identity map to the space $\boldsymbol{F}_1$. But then, since $\boldsymbol{F}_1$ has a dimension 2 and G_1 is skew-symmetric over $\boldsymbol{F}_1$, it is clear that for all orthonormal basis $\mathbf{f}_1, \mathbf{f}_2$ of $\boldsymbol{F}_1$, it is possible to write:

$$\exists \eta_1 \in \mathbb{R}, \ \ \forall \mathbf{x} \in \boldsymbol{F}_1, \ \ G_1\,(\mathbf{x}) = \eta_1 \mathbf{f}_1 \wedge \mathbf{f}_2 : \mathbf{x}, \tag{11.77}$$

then, it is immediately obtained that:

$$\exists \eta_1 \in \mathbb{R}, \ \ \forall \mathbf{x} \in \boldsymbol{F}_1, \ \ G_1^2\,(\mathbf{x}) = -\eta_1^2 \mathbf{x} = \lambda_1 \mathbf{x}. \tag{11.78}$$

4) Within a Euclidean space, it is always possible to write that $Im\,(g) = \boldsymbol{F}_1 \oplus^{\perp} \boldsymbol{F}_1^{\perp}$. Now, since $\boldsymbol{F}_1$ is stable with respect to G, $\boldsymbol{F}_1^{\perp}$ is stable with respect to G. We can now consider the restriction of G to $\boldsymbol{F}_1^{\perp}$. This restriction admits exactly the same hypotheses as the function G over $Im\,(g)$. By induction, and since we already know that there exists p for which $\dim\,(Im\,(g)) = 2p$, we show the existence of p subspaces with a dimension 2, such that:

$$Im\,(g) = \boldsymbol{F}_1 \oplus^{\perp} \boldsymbol{F}_2 \oplus^{\perp} \cdots \oplus^{\perp} \boldsymbol{F}_p, \tag{11.79}$$

so that G is stable over each $\boldsymbol{F}_i$ and that these G^2, restricted to $\boldsymbol{F}_i$, are equal to $\lambda_i \mathbb{I}_i$, where $\mathbb{I}_i$ represents the restriction to $\boldsymbol{F}_i$ of the map G. On the other hand, in order to find an orthodecomposition of G, it is enough to find two orthonormal vectors $\mathbf{f}_{2i-1}, \mathbf{f}_{2i}$ for each $\boldsymbol{F}_i$. We then have:

$$\mathbf{G} = \sum_{i=1}^{i=p} \eta_i \mathbf{f}_{2i-1} \wedge \mathbf{f}_{2i}, \ \ -\eta_i^2 = \lambda_i. \tag{11.80}$$

REMARK 11.3.– Before proceeding further, note some important points:

1) It is evident that in the pseudo-Euclidean case, the situation seems more complicated *a priori*: the inductive reasoning applied before cannot be guaranteed to work anymore, since the subspaces of a pseudo-Euclidean space are not necessarily pseudo-Euclidean themselves. On the other hand, a space not necessarily can be written as a direct sum with its orthogonal space.

2) Actually, all these problems are mainly due to the isotropic cone, i.e. the set of vectors that satisfy $q(\mathbf{x}) = 0$. In the absence of a reciprocal proposition of the diagonalization for the pseudo-orthodecomposition in a general case, we shall restrict ourselves, in a first phase, to the Minkowski's case, in other words, to pseudo-Euclidean spaces with a signature $s = 1$. In this way, according to theorem 6.8, the isotropic cone cannot contain a linear space with a dimension greater than 1. Once this space is isolated, the situation becomes Euclidean again, and the previous demonstration is still valid. ■

THEOREM 11.9 (reciprocal, Minkowski's case).– Let $(\boldsymbol{E}, b)$ be a ***Minkowski*** space (i.e. pseudo-Euclidean with a signature $s = 1$) and $\mathbf{E}$ be an element of $\Lambda^2\boldsymbol{E}$. Suppose that the endomorphism $g := \gamma^{-1}(\mathbf{E})$ is such that g^2 restricted to $\left(Im(g), b_{Im(g)}\right)$ is diagonalizable. Then, $\mathbf{E}$ is pseudo-orthodecomposable.

As in the previous case, G will denote the restriction of g to $Im(g)$:

1) First, we have that:

LEMMA 11.3.– The endomorphism G satisfies $\dim(\ker G) \leq 1$.

PROOF.– In fact, we have $G(\mathbf{x}) = 0$ if and only if $\mathbf{x} \in Im(g) \bigcap \ker(g)$. Since g is skew-symmetric, according to proposition 11.3, it results that $\ker(g) = [Im(g)]^{\perp}$. We can thus write that:

$$G(\mathbf{x}) = \mathbf{0} \Leftrightarrow \mathbf{x} \in [Im(g)] \bigcap [Im(g)]^{\perp}. \quad [11.81]$$

Now, any $\mathbf{x}$ of the latter set necessarily satisfies $q(\mathbf{x}) = 0$. By theorem 6.8, the dimension of the greatest isotropic subspace equals $\min(s, n-s)$, which here is 1 since the signature of the pseudo-dot product equals $s = 1$.

2) A clarification which will be useful:

LEMMA 11.4.– Let $(\boldsymbol{F}, b)$ be such that b is definite. Then, for all subspace $\boldsymbol{G}$ included in $\boldsymbol{F}$, we have $\boldsymbol{F} = \boldsymbol{G} \oplus^{\perp} \boldsymbol{G}^{\perp}$.

PROOF.– Let us recall that definite implies nondegenerate (see proposition 6.3). Consequently, we always have $\dim(\boldsymbol{G}) + \dim(\boldsymbol{G}^{\perp}) = \dim(\boldsymbol{F})$ (see proposition 6.5). On the other hand, if $\mathbf{x} \in \boldsymbol{G} \bigcap \boldsymbol{G}^{\perp}$ then $q(\mathbf{x}) = 0$ and thus $\mathbf{x} = \mathbf{0}$ since q is definite. Thus, we indeed have the stated result.

3) Let us consider a slightly longer lemma:

LEMMA 11.5.– Let H be a skew-symmetric endomorphism of a space $\boldsymbol{F}$ equipped with a definite symmetric bilinear form b. Suppose that H^2 is diagonalizable and invertible. Then, the dimension of $\boldsymbol{F}$ is even.

Note that in pseudo-Euclidean spaces, symmetry does not necessarily entail the diagonalization and thus *a priori*, although it is symmetric, H^2 might not be diagonalizable.

PROOF.– Since H^2 is diagonalizable, let $\mathbf{e}_1 \neq \mathbf{0}$ be such that $H^2(\mathbf{e}_1) = \lambda \mathbf{e}_1$. We know that $\lambda \neq 0$ since H^2 is invertible. Let us set $\mathbf{e}_2 = H(\mathbf{e}_1)$. Then, it is immediate that $H^2(\mathbf{e}_2) = \lambda \mathbf{e}_2$. On the other hand, we have:

$$b\left(\mathbf{e}_2, H^2(\mathbf{e}_2)\right) = -b\left(H(\mathbf{e}_2), H(\mathbf{e}_2)\right) = -b(\lambda \mathbf{e}_1, \lambda \mathbf{e}_1) = -\lambda^2 q(\mathbf{e}_1) \neq 0. \qquad [11.82]$$

In particular, $\mathbf{e}_2$ is non-null. Furthermore, we have $b(\mathbf{e}_1, \mathbf{e}_2) = b(H(\mathbf{e}_1), \mathbf{e}_1) = 0$ since H is skew-symmetric. This implies that the family $\mathbf{e}_1, \mathbf{e}_2$ is orthogonal. Since $\mathbf{e}_1, \mathbf{e}_2$ are non-null, this additionally implies that this family is linearly independent, due to the fact that:

$$\alpha_1 \mathbf{e}_1 + \alpha_2 \mathbf{e}_2 = \mathbf{0} \Rightarrow \alpha_1 q(\mathbf{e}_1) = 0 \text{ and } \alpha_2 q(\mathbf{e}_2) = 0 \Rightarrow \alpha_1 = 0 \text{ and } \alpha_2 = 0, \qquad [11.83]$$

because $q(\mathbf{e}_1) \neq 0$ and $q(\mathbf{e}_2) \neq 0$, as q is definite and the vectors are non-null. Consider $\boldsymbol{G}_1 := \langle \mathbf{e}_1, \mathbf{e}_2 \rangle$ the space generated by $\mathbf{e}_1, \mathbf{e}_2$. Since we have:

$$H(\mathbf{e}_1) = \mathbf{e}_2, \quad H(\mathbf{e}_2) = H^2(\mathbf{e}_1) = \lambda \mathbf{e}_1, \qquad [11.84]$$

this entails that $\boldsymbol{G}_1$ is an eigensubspace of H^2 (thus stable with respect to H^2) which is also stable with respect to H. Consider $\boldsymbol{F}_1 := \boldsymbol{F} \ominus^{\perp} \boldsymbol{G}_1$. Then, $\boldsymbol{F}_1$ is stable with respect to H, H^2 since H and H^2 are respectively pseudo-skew-symmetric and pseudo-symmetric; therefore, if they are stable over a subspace, they are also stable over the subspace's orthogonal (see proposition 11.4). Let us define H_1 as the restriction of H to $\boldsymbol{F}_1$. Then, H_1^2 and H_1 satisfy the same hypotheses over $\boldsymbol{F}_1$ as for H, H^2 over $\boldsymbol{F}$. We can then find, in $\boldsymbol{F}_1$, a subspace $\boldsymbol{G}_2$ with a dimension two that is an eigensubspace for H_1^2 and stable for H_1. We consider then $\boldsymbol{F}_2 := \boldsymbol{F}_1 \ominus^{\perp} \boldsymbol{G}_2$. Let us denote by k the floor function of $\dim(\boldsymbol{F})/2$. By induction, it is possible to find some $\boldsymbol{G}_i, i \in [1, k]$ such that:

$$\boldsymbol{G}_1 \oplus^{\perp} \cdots \oplus^{\perp} \boldsymbol{G}_k \subset \boldsymbol{F} \qquad [11.85]$$

with H^2 and H stable over each of these subspaces $\boldsymbol{G}_i$. Then, if $\dim(\boldsymbol{F}) = 2k+1$, there remains an isolated non-null vector that is orthogonal to all the spaces $\boldsymbol{G}_i$; let us denote it by $\mathbf{e}_{2k+1}$. However, since H, H^2 are stable over $\boldsymbol{G}_i$, it is clear that they are also stable over the space generated by $\mathbf{e}_{2k+1}$. We thus have $H(\mathbf{e}_{2k+1}) = \alpha \mathbf{e}_{2k+1}$. Now, $b(H(\mathbf{e}_{2k+1}), \mathbf{e}_{2k+1}) = 0$ (since H is skew-symmetric) and thus $\alpha q(\mathbf{e}_{2k+1}) = 0$, and finally $\alpha = 0$ since $\mathbf{e}_{2k+1} \neq \mathbf{0}$ and q is definite. But then, $H(\mathbf{e}_{2k+1}) = \mathbf{0}$ with $\mathbf{e}_{2k+1} \neq \mathbf{0}$, which contradicts the hypothesis that H^2 is invertible.

4) An important implication of the hypothesis is that $s \in \{0,1\}$ and that G is diagonalizable:

LEMMA 11.6.– We have $Im(g) = \ker(G^2) \oplus^{\perp} Im(G^2)$, and if $\dim(\ker G) = 1$, the subspace $Im(G^2)$ does not contain any non-null isotropic vector.

PROOF.– The property $Im(g) = \ker(G^2) \oplus^{\perp} Im(G^2)$ simply results from the hypothesis of diagonalization over G^2. In particular, all the elements of $Im(G^2)$ are orthogonal to those of $\ker(G^2)$. Suppose that $\dim(\ker G) = 1$ with $\mathbf{e}_1$ being an isotropic vector and basis of $\ker(G)$. Since $\ker(G) \subset \ker(G^2)$, we know that $\ker(G^2)$ contains an isotropic vector $\mathbf{e}_1$. If $Im(G^2)$ also contains a non-null isotropic vector, denoted by $\mathbf{e}_2$, then $\mathbf{e}_1$ is necessarily orthogonal to $\mathbf{e}_2$. By employing:

$$q(\alpha \mathbf{e}_1 + \beta \mathbf{e}_2) = \alpha^2 q(\mathbf{e}_1) + \beta^2 q(\mathbf{e}_2) + 2\alpha\beta b(\mathbf{e}_1, \mathbf{e}_2), \qquad [11.86]$$

we obtain the fact that we are able to construct a space with a dimension two that is isotropic: this contradicts the fact that the dimension of such an isotropic space is at most 1, since $s \in \{0,1\}$.

5) We therefore deduce a corollary:

COROLLARY 11.5.– The dimension of $Im(G^2)$ is even.

PROOF.–

a) If $\ker G = \{\mathbf{0}\}$, then G^2 is an isomorphism, and therefore $Im(G^2) = Im(g)$, which has an even dimension since g is a pseudo-skew-symmetric endomorphism associated with $\mathbf{E}$.

b) If $\dim(\ker G) = 1$, then according to the previous lemma 11.6, $Im(G^2)$ does not contain isotropic elements. In particular, the restriction of the form q to $Im(G^2)$ is definite. Furthermore, thanks to the diagonalization of G^2, we have:

$$Im(g) = \ker(G^2) \oplus^{\perp} Im(G^2), \qquad [11.87]$$

the restriction of G^2 to the space $Im(G^2)$ is invertible and diagonalizable, and the form q is definite over $Im(G^2)$. Using the preliminary lemma 11.5, we deduce that $Im(G^2)$ has an even dimension.

6) A short lemma:

LEMMA 11.7.– If $\dim(\ker(G)) = 1$, then $1 \leq \dim(\ker(G^2)) \leq 2$.

PROOF.– Since we always have $\ker(G) \subset \ker(G^2)$, the minoration of the dimension is straightforward. Let $\mathbf{e}_1$ be a basis of $\ker(G)$. Let us now solve:

$$G^2(\mathbf{x}) = \mathbf{0} \Leftrightarrow \exists\alpha,\ G(\mathbf{x}) = \alpha\mathbf{e}_1. \qquad [11.88]$$

Suppose that there exists $\mathbf{e}_2$ such that $G(\mathbf{e}_2) = \mathbf{e}_1$, then we can perform the previous solving procedure by writing that:

$$G(\mathbf{x}) = \alpha\mathbf{e}_1 \Leftrightarrow g(\mathbf{x} - \alpha\mathbf{e}_2) = \mathbf{0} \Leftrightarrow \mathbf{x} - \alpha\mathbf{e}_2 = \beta\mathbf{e}_1, \qquad [11.89]$$

which shows that the dimension of $\ker(G^2)$ is at most two. Now, if the equation $G(\mathbf{y}) = \mathbf{e}_1$ does not admit a solution, then the only possibility of having $G(\mathbf{x}) = \alpha\mathbf{e}_1$ is that $\alpha = 0$. We then deduce that $\mathbf{x} = \beta\mathbf{e}_1$ and the dimension of the kernel is at least one.

7) A corollary follows:

COROLLARY 11.6.– If $\dim(\ker(G^2)) \neq 0$, then $\dim(\ker(G^2)) = 2$ and $Im(G^2)$ does not contain any isotropic vector.

PROOF.– Actually, if $\dim(\ker(G^2)) \neq 0$, this means, by the previous lemma, that $1 \leq \dim(\ker(G^2)) \leq 2$. Now, thanks to the diagonalization of G^2, we know that:

$$\dim(Im(g)) = \dim(\ker(G^2)) + \dim(Im(G^2)), \qquad [11.90]$$

and furthermore we know that $\dim(Im(G^2))$ is even and $\dim(Im(g))$ is also even. Hence we deduce that $\dim(\ker(G^2))$ is even. By adding the constraint of the previous corollary, we finally have $\dim(\ker(G^2)) = 2$.

In any case, we can write that:

LEMMA 11.8.– The two followings statements are true:

1) the space $Im(G^2)$ is nondegenerate. In particular, the space $(Im(G^2), b)$ is pseudo-Euclidean;

2) the endomorphism G^2 restricted to $Im(G^2)$ is invertible (and diagonalizable).

PROOF.– Let us illustrate a proof of these two statements:

1) In fact, if $\dim\left[\ker\left(G^2\right)\right] = 2$, we have seen that the space $Im\left(G^2\right)$ is defined for the form q, and, in particular, it is nondegenerate (see the implication of proposition 6.3) for b. If $\ker(G) = \{0\}$, then we necessarily have that $Im(g) \bigcap \ker(g) = \{0\}$ and since g is skew-symmetric, we have $\ker(g) = (Im(g))^{\perp}$ which implies that $Im(g) \bigcap Im(g)^{\perp} = \{\mathbf{0}\}$ and thus $Im(g)$ is nondegenerate (see the characterization of corollary 6.3). Now, if G is invertible, we have $Im\left(G^2\right) = Im(g)$ and thus $Im\left(G^2\right)$ is indeed nondegenerate for the form b.

2) The second statement is based only on the diagonalization of G^2, since we actually have $Im(g) = \ker\left(G^2\right) \oplus^{\perp} Im\left(G^2\right)$ and thus if $\mathbf{x}$ belongs to $Im\left(G^2\right)$, then it satisfies $G^2(\mathbf{x}) = \mathbf{0}$ only if it is null.

PROPOSITION 11.6.– Let $(\boldsymbol{F}, b)$ be a pseudo-Euclidean space, with a dimension $2p$, and let G be a skew-symmetric endomorphism, invertible over $\boldsymbol{F}$, such that G^2 (which is thus invertible) is diagonalizable. We suppose that the dimension of the greatest isotropic linear space is equal to 1 (Minkowski's hypothesis). Then, there exist $\boldsymbol{G}_1, \cdots, \boldsymbol{G}_p$ subspaces with a dimension two such that:

1) we have: $\boldsymbol{F} = \boldsymbol{G}_1 \oplus^{\perp} \cdots \oplus^{\perp} \boldsymbol{G}_p$;

2) each $\boldsymbol{G}_k$ is an eigensubspace of G^2 for an eigenvalue λ_k. Furthermore, it is stable with respect to G;

3) the restriction of G to each $\boldsymbol{G}_k$ can be written in the form $\sqrt{|\lambda_k|}\mathbf{f}_{2k-1} \wedge \mathbf{f}_{2k}$ where $\mathbf{f}_{2k-1}, \mathbf{f}_{2k}$ is a pseudo-orthonormal basis of $\boldsymbol{G}_k$.

PROOF.– Consider $\mathbf{e}_1$ an eigenvector of G^2. We can always construct an orthogonal basis of $\boldsymbol{F}$ that contains $\mathbf{e}_1$ and that is an eigenvector of G^2 (this results from the diagonalization hypothesis). In particular, since space $\boldsymbol{F}$ is nondegenerate, this implies that it is impossible that $q(\mathbf{e}_1) = 0$ since otherwise vector $\mathbf{e}_1$ would be such that:

$$\forall \mathbf{x} \in \boldsymbol{F}, \; b(\mathbf{e}_1, \mathbf{x}) = 0, \qquad [11.91]$$

and thus form b would be degenerate. Let us set $G(\mathbf{e}_1) = \mathbf{e}_2$. Clearly, $\mathbf{e}_2$ is not null since G^2 is invertible over $\boldsymbol{F}$. Moreover, due to skew symmetry, the vectors $\mathbf{e}_1, \mathbf{e}_2$ are orthogonal. Finally, they form an independent family since:

$$\alpha \mathbf{e}_1 + \beta \mathbf{e}_2 = \mathbf{0} \Rightarrow \alpha q(\mathbf{e}_1) = 0 \Rightarrow \alpha = 0 \Rightarrow \beta = 0, \qquad [11.92]$$

the latter implication being due to the fact that $\mathbf{e}_2$ is not null. The space generated by $\mathbf{e}_1, \mathbf{e}_2$ is stable with respect to G since $\mathbf{e}_1$ is an eigenvector for G^2. Finally, it is clear that G^2 restricted to $\boldsymbol{G}_1 := \langle \mathbf{e}_1, \mathbf{e}_2 \rangle$ is proportional to $\lambda_1 \mathbb{I}_1$, with $\mathbb{I}_1$ the restriction to $\boldsymbol{G}_1$ of the identity. Since $\mathbf{e}_2$ is an eigenvector of G^2, and since G^2 is diagonalizable in $Im\left(G^2\right)$, it is impossible that $q\left(\mathbf{e}_2\right) = 0$, as otherwise we would be able to construct an orthogonal basis of $Im\left(G^2\right)$ with an isotropic vector, which is impossible since $Im\left(G^2\right), b$ is pseudo-Euclidean. Therefore, the space $\boldsymbol{G}_1$ is nondegenerate and consequently we indeed have:

$$Im\left(G^2\right) = \boldsymbol{G}_1 \oplus^{\perp} \boldsymbol{G}_1^{\perp}, \qquad [11.93]$$

the endomorphisms G, G^2 are stable over $\boldsymbol{G}_1^{\perp}$, and the restriction of G^2 to $\boldsymbol{G}_1^{\perp}$ is also diagonalizable and invertible. Since $\boldsymbol{G}_1$ is nondegenerate, the space $\boldsymbol{G}_1, b$ is pseudo-Euclidean. By induction, we obtain the decomposition stated in the proposition.

We finally conclude the reciprocal statement of the Minkowski's case in the following way: if $\ker(G) \neq 0$, $\ker\left(G^2\right)$ has a dimension two. The restriction of G to $\ker\left(G^2\right)$ is such that $\gamma^{-1}(G)$ restricted to $\ker\left(G^2\right)$ is pseudo- orthodecomposable (since it has a dimension two, and this is a direct consequence of the Minkowki's hypothesis). Next, we consider the restriction of G^2 to $Im\left(G^2\right)$. Thanks to the previous proposition, the restriction of $\gamma^{-1}(G)$ to each $\boldsymbol{G}_i$ is pseudo-orthodecomposable. Since the sum of the G_i is a direct orthogonal sum, and since the latter are themselves in a direct orthogonal sum with $\ker\left(G^2\right)$, we finally have the pseudo-orthodecomposability of $\gamma^{-1}(G)$.

Bibliography

[CAU 82] CAUCHY A.-L., *Oeuvres complètes d'Augustin Cauchy. Série 1, Tome 11*, Gauthier-Villars et fils, Paris, 1882.

[DIE 79] DIEUDONNÉ J., "The Tragedy Of Grassmann", *Linear and Multi-linear Algebra*, vol. 8, pp. 1–14, 1979.

[ENG 11] ENGEL F., *Gesammelte mathematische und physikalische werke. Dritten bandes zweiter teil: Grassmanns Leben*, vol. 3, B.G. Teubner, Leipzig, 1911.

[FEA 79] FEARNLEY-SANDER D., "Hermann Grassmann and the creation of linear algebra", *American Mathematical Monthly*, vol. 86, pp. 809–817, 1979.

[FLA 05] FLAMENT D., "H. G. Grassmann et l'introduction d'une nouvelle discipline mathématique: l'Ausdehnungslehre", *Philosophia Scientia: Fonder autrement les mathématiques*, vol. 5, pp. 81–141, 2005.

[GRA 44] GRASSMANN H., *Die lineale Ausdehnungslehre*, Wiegand, Leipzig, 1844.

[GRA 62] GRASSMANN H., *Die Ausdehnungslehre, vollstandig und in strenger Form bearbeitet*, Enslin, Berlin, 1862.

[PET 11] PETSCHE H.-J., LEWIS A.C., LIESEN J., *et al.* (eds.), *From Past to Future: Grassmann's Work in Context*, Birkhäuser, Basel, 2011.

[SCH 81] SCHWARTZ L., *Les tenseurs*, Hermann, Paris, 1981.

[SCH 96] SCHUBRING G. ed., *Hermann Günther Grassmann (1809–1877): Visionary Mathematician, Scientist and Neohumanist Scholar*, Springer, Dordrecht, 1996.

[SCH 97] SCHWARTZ L., *Un mathématicien aux prises avec le siècle*, Odile Jacob, Paris, 1997.

Index

H, I, L

M, N, P

R, S, T, U

Printed in the United States
By Bookmasters